P • O • C • K • E • T • S

EARTH FACTS

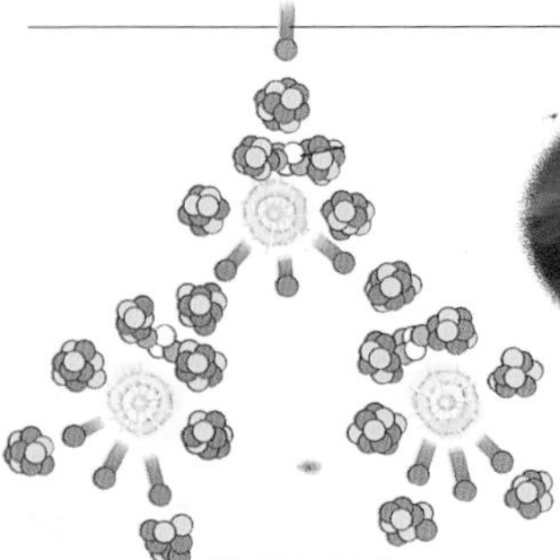

Nuclear fission

Island arc

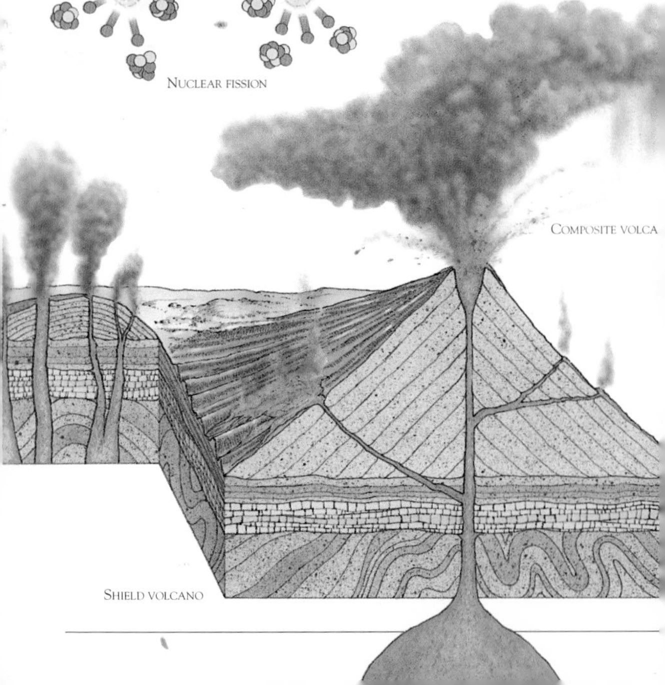

Composite volca

Shield volcano

P • O • C • K • E • T • S

EARTH FACTS

Written by
CALLY HALL and
SCARLETT O'HARA

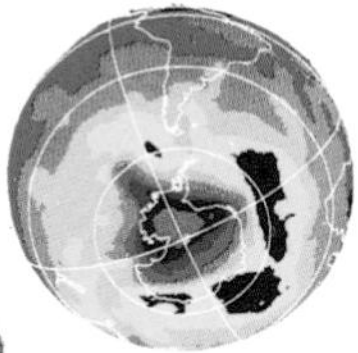

SATELLITE PICTURE OF EARTH'S OZONE HOLE

ONION-LAYERING

CARBONIFEROUS FOREST

DORLING KINDERSLEY
London • New York • Stuttgart

A DORLING KINDERSLEY BOOK

Project editor Scarlett O'Hara
Art editor Susan Downing
Senior editor Laura Buller
Senior art editor Helen Senior
Editorial consultant Cally Hall
Picture research Charlotte Bush
Christine Rista
Production Louise Barratt
US editor Jill Hamilton
US consultant Warren Yasso, Ph.D.
Columbia University

First American Edition, 1995
2 4 6 8 10 9 7 5 3 1
Published in the United States by
Dorling Kindersley Publishing, Inc.,
95 Madison Avenue,
New York, New York 10016

 Published in Great Britain by Dorling Kindersley Limited.
Distributed by Houghton Mifflin Company, Boston.

Library of Congress Cataloging-in-Publication Data
Earth facts / Cally Hall and Scarlett O'Hara
p. cm. – (A DK pocket)
Includes index.
ISBN 1-56458-891-2
1.Earth sciences – Juvenile literature. [1. Earth sciences]
I. Hall, Cally. II .Series.
QE29.E25 1995
550–dc20
94–32620
CIP

Color reproduction by Colourscan, Singapore
Printed and bound in Italy by L.E.G.O.

CONTENTS

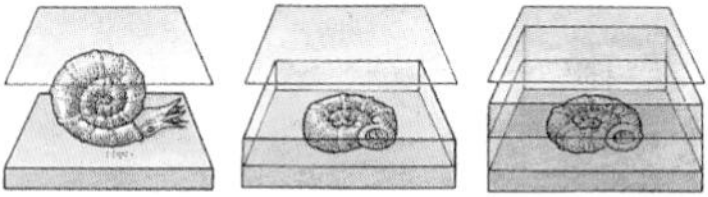

How to use this book

These pages show you how to use *Pockets: Earth Fact* The book is divided into 13 sections. Each section contains information on one aspect of the Earth. Th pages in the section give further details on the topic At the beginning of each section there is a guide to the contents of that section.

Corner coding

In the corner of each page a coloured square indicates the section's topic.

- Planet Earth
- Earth's plates and continents
- Volcanoes
- Earthquakes
- Landscape, weathering, and erosion
- Rocks and minerals
- Mineral resources
- Mountains, valleys, and caves
- Glaciation
- Oceans, islands, and coasts
- Rivers and lakes
- Climate and weather
- A future for the Earth

Corner coding

Heading

Introduction

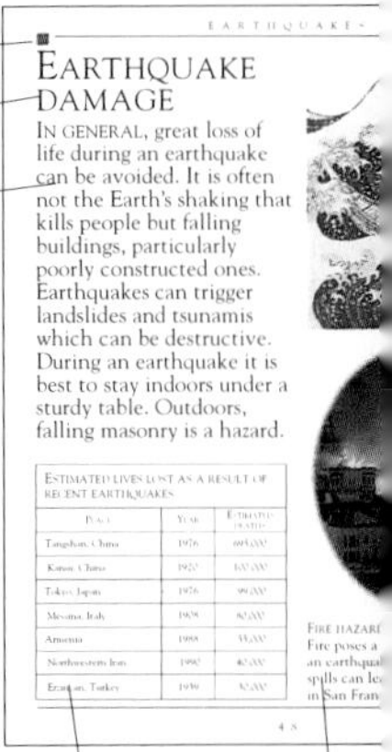

Earthquakes
Earthquake damage
In general, great loss of life during an earthquake can be avoided. It is often not the Earth's shaking that kills people but falling buildings, particularly poorly constructed ones. Earthquakes can trigger landslides and tsunamis which can be destructive. During an earthquake it is best to stay indoors under a sturdy table. Outdoors, falling masonry is a hazard.
Estimated lives lost as a result of recent earthquakes
Fire hazar
Fire poses a
an earthqua
spills can le
in San Fran

Chart

Cap

Heading and introduction

The subject of the page is indicated here. Within the section, each new page begins a new aspect of the subject.

Charts

Many pages contain charts. These give information on the world's highest, longest, tallest, deepest, etc. This chart indicates the number of lives lost in several serious earthquakes.

Labels

Some pictures have labels. These give extra informatic or identify a picture if it is not immediately obvious from the text.

ᴺNNING HEADS
a reminder of the :ion, the left-hand e has a running d with the section ıe.The running d of the right-hand e gives the subject ıe particular page.

FACT BOXES
These at-a-glance information boxes appear on many pages. They provide fascinating details about the subject. This fact box has details on tsunamis.

FORMATION BOXES
To explain a process, a formation diagram may be used. These show several stages in a process and may have annotation. The diagrams and their accompanying captions are enclosed in a box.

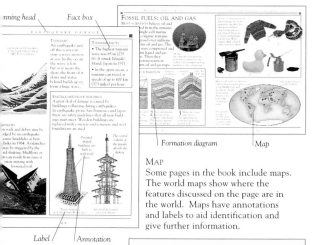

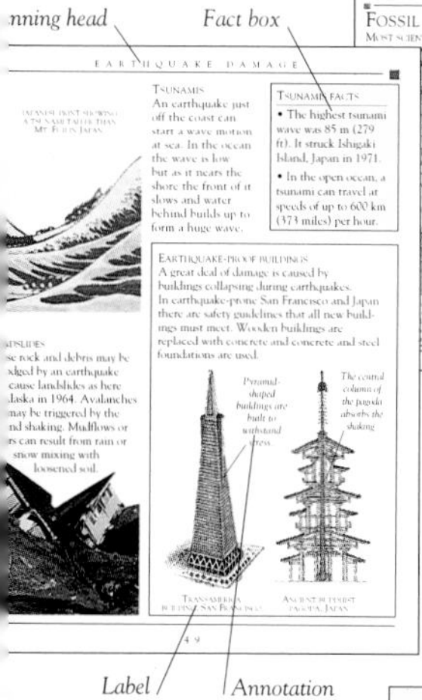

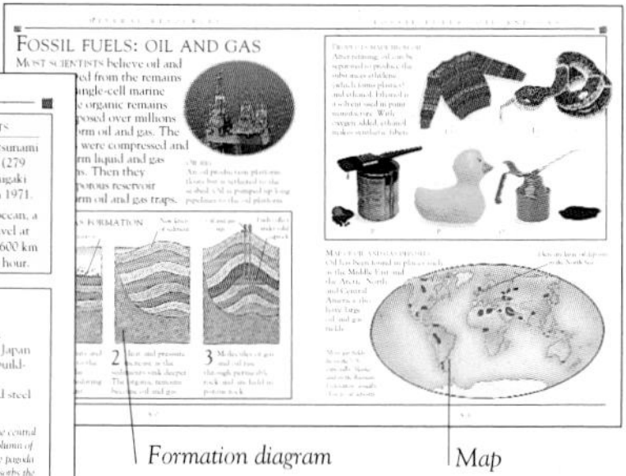

MAP
Some pages in the book include maps. The world maps show where the features discussed on the page are in the world. Maps have annotations and labels to aid identification and give further information.

ᴾTIONS AND ANNOTATIONS
h illustration is accompanied caption. Annotations, in :s, point out the features of llustration or diagram and ılly have leader lines.

INDEX
At the back of the book, there is an index. It lists alphabetically every subject included in the book. By referring to the index, information on particular topics can be found quickly.

PLANET EARTH

HOW THE EARTH WAS FORMEI

ABOUT 5 BILLION years ago our Solar System began to take shape. The Sun and the nine planets formed from a cloud of dust and gas swirling in space. Some scientists believe that the center of this cloud coole and contracted to form the Sun. Gravity pulled the planets from the rest of the cloud. Other scientists suggest that the dust cloud formed asteroids that joined together tc make the Sun and plane

A dense atmosphere of cosmic gases surrounded the Earth.

1 FORMING THE SUN
A spinning cloud of gas and dust contracted to form the Sun. Cooler matter from this dust cloud combined to shape the planets.

2 FORMING THE EARTH
The Earth's radioactivity caused the surface to melt. Lighter minerals floated to the surface and heavier elements, such as iron and nickel, sank to form the Earth's core.

3 THE EARTH'S CRUST
About 4 billion years ago, the Earth's crust began to form. Blocks of cooling, solid rock floated on a molten rock layer. The rock sometimes sank and remelted before rising again.

.RTH FACTS

- The Earth orbits the
n at 18.5 miles/sec
).8 km/sec).
- Oceans cover 70.8%
the Earth's surface.
- Earth is not a sphere –
oulges in the middle.
- The Earth completes
urn on its axis every
hours, 56 minutes.

COMPOSITION OF THE EARTH

The elements here are divided by weight. Earth's crust consists mostly of oxygen, silicon, and aluminum. Heavier metals such as iron and nickel are found in the core.

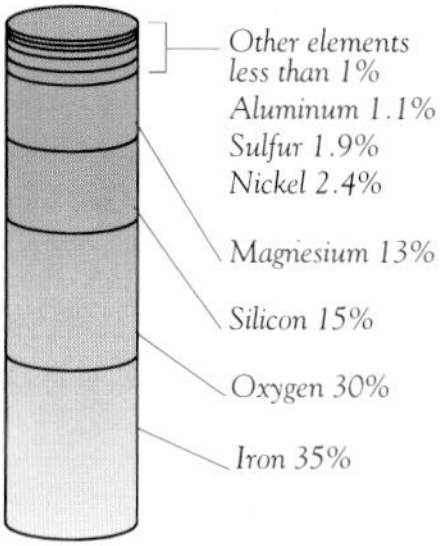

4 MAKING THE ATMOSPHERE
The Earth's crust thickened. It took several million years for volcanic gases to form the atmosphere. Water vapor condensed to make oceans.

6 THE EARTH TODAY
Earth's unique conditions are just right to support a variety of life. Our planet, though, continues to change. Tectonic plates are moving, pulling some continents nearer and pushing others farther apart.

LAND
FORMS
ut 3.5
on years ago
ocontinents formed on the crust.
ay's continents look very different.

THE EARTH IN SPACE

EARTH IS A DENSE rocky planet, third nearest to th Sun, and tiny compared with Jupiter and Saturn While Earth rotates on its axis once each day, i also orbits the Sun once each year, held in orb by the Sun's gravity. One moon revolves arou the Earth. From space the Earth looks blue a calm but under its oceans, deep beneath th crust, the Earth's core is fiery and white-h

MERCURY
- 87.96 days to orbit Sun
- diameter 3,031 miles (4,878 km)

EARTH
- 365.26 days to orbit Sun
- diameter 7,926 miles (12,756 km)
- 1 moon

MARS
- 686.98 days to orbit Sun
- diameter 4,217 miles (6,786 km)
- 2 moons

VENUS
- 224.7 days to orbit Sun
- diameter 7,520 miles (12,102 km)

SUN
- diameter 865,000 miles (1,391,980 km)

THE SOLAR SYSTEM
Our Solar System consists of ni planets, as well as moons, aster comets, meteorites, dust, and g of these orbit a central star – th

The Great Red Spot is a cyclone.

JUPITER
- 11.86 years to orbit Sun
- diameter 88,846 miles (142,984 km)
- 16 moons
- 1 ring

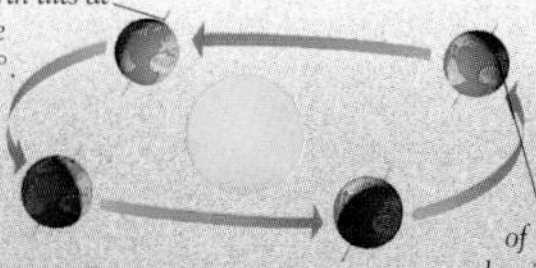
ne Earth tilts at
: angle
23.5°.
This side
of the planet
has its winter.

ARTH'S ORBIT

.s the Earth turns on its axis, it also orbits the un. When the Northern Hemisphere faces ie Sun it has its summer. At the same time ie Southern Hemisphere faces away from the un and has its winter. The equator faces owards the Sun most of the time and there e no significant seasonal changes there.

DISTANCE FROM THE SUN		
PLANET	MILLION KM	MILLION MILES
Mercury	58	36
Venus	108	67
Earth	150	93
Mars	228	142
Jupiter	778	483
Saturn	1,427	887
Uranus	2,871	1,784
Neptune	4,497	2,794
Pluto	5,914	3,675

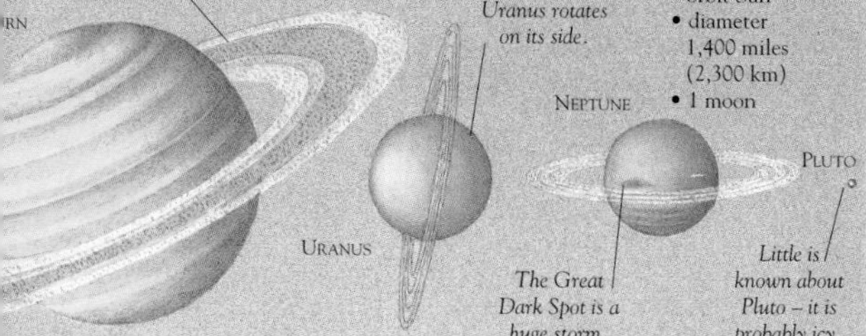
Saturn's rings are
made of pieces of icy
rock and dust.
RN
Uranus rotates
on its side.
NEPTUNE
URANUS
PLUTO
The Great
Dark Spot is a
huge storm.
Little is
known about
Pluto – it is
probably icy.

PLUTO
- 248.54 years to orbit Sun
- diameter 1,400 miles (2,300 km)
- 1 moon

ATURN
29.46 years to orbit Sun
diameter 74,898 miles (120,536 km)
18 moons
7 rings

URANUS
- 84 years to orbit Sun
- diameter 31,763 miles (51,118 km)
- 15 moons
- 11 rings

NEPTUNE
- 164.79 years to orbit Sun
- diameter 30,775 miles (49,528 km)
- 8 moons
- 4 rings

EARTH'S MAGNETIC FIELD

THE EARTH BEHAVES like a giant magnet. Molten nickel and iron flowing in the molten outer core of the Earth produce an electric current. This electricity creates a magnetic field, or magnetosphe that extends into space. Like a magnet, the Earth h two magnetic poles. From time to time, the magnet poles reverse polarity. The last time they changed w about 700,000 years ago. No one knows why this happens.

MAGNETIC POLES
North and south geographical poles lie at either end of the Earth's axis (the invisible line around which the Earth turns). The magnetic poles' position varies over time. It is the Earth's magnetic field that causes a compass needle to point north.

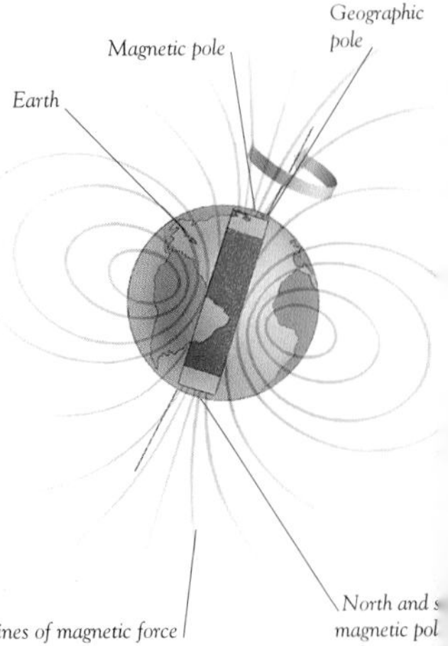

Invisible lines of magnetic force form a pattern around the Earth. Closer lines reveal a stronger magnetic field.

North and s magnetic pol always nec north and s geographical

MAGNETIC FACTS

- Whales and birds use the Earth's magnetic field to help them navigate.
- Every second the Sun sheds at least a million tons (tonnes) of matter into the solar wind.

ɜNETOSPHERE

h's magnetosphere extends ıt 37,000 miles (60,000 km) space. It protects the Earth ı some of the Sun's most nful energetic particles.

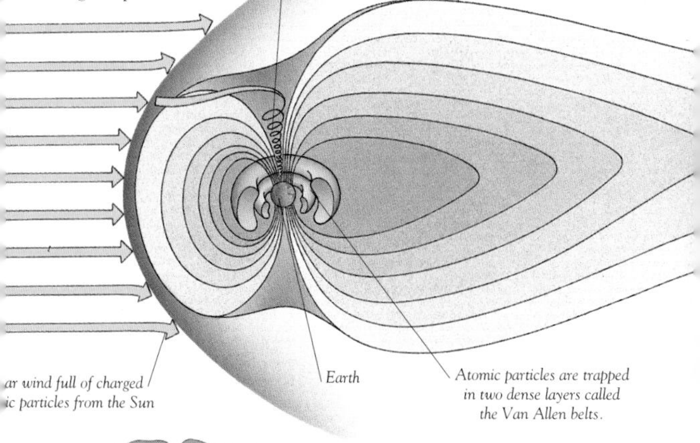

Solar wind particles caught by Earth's atmosphere glow as the auroras.
Electrically charged particles from the Sun push the magnetosphere out of shape.
ar wind full of charged ic particles from the Sun
Earth
Atomic particles are trapped in two dense layers called the Van Allen belts.

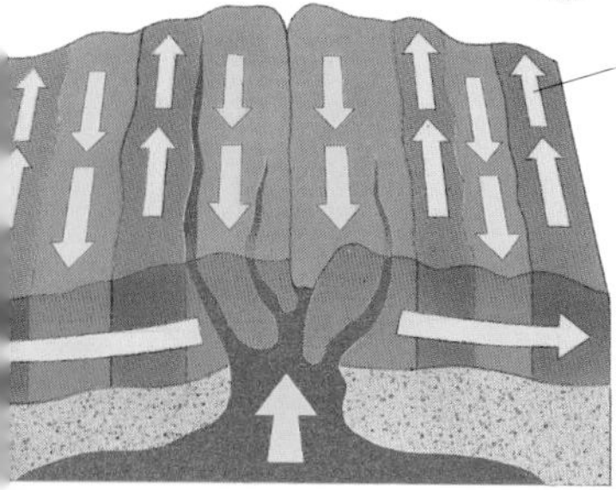

The polarity of new crust reverses between north and south.

MAGNETIC CRUST

New oceanic crust rises out of the Earth at midoceanic ridges. As the rock solidifies, a record of the Earth's magnetism is locked into it. Earth's changing polarities lead to a magnetic pattern in the rock, symmetrical either side of the spreading ridge.

EARTH'S ATMOSPHERE

THE EARTH IS WRAPPED in a blanket of gases called t atmosphere. This thin layer protects the Earth from the Sun's fierce rays and from the hostile conditions of outer space. There are five layers in the Earth's atmosphere before the air merges with outer space. The lowest layer holds air and water vapor that support life, and our weather and climate.

A THIN LAYER
The Earth's atmosphere is actually a thin band around the Earth. If the Earth were an orange, the atmosphere would be as thin as the skin of the orange.

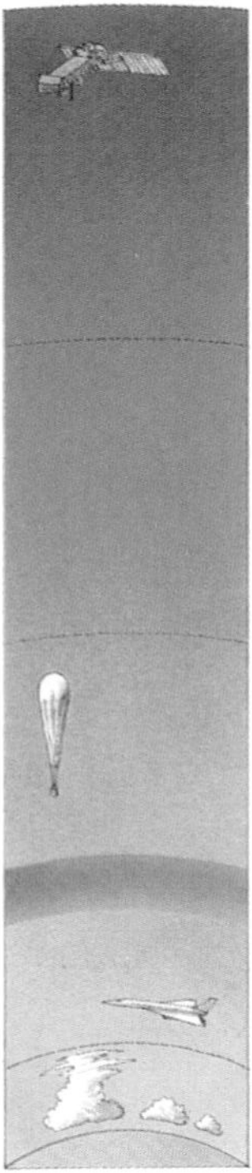

EXOSPHERE
- begins at 560 miles (900 km)
- thin layer before spacec reach outer space

THERMOSPHERE
- 50-280 miles (80-450 k
- reaches 3,600°F (2,000°C)
- contains the ionospher electrically charged air reflects radio waves

MESOSPHERE
- 30-50 miles (50-80 km)
- meteors burn up and cause shooting stars

STRATOSPHERE
- 12-30 miles (20-50 km)
- ranges from -76°F (-60° to just about freezing po at the top
- calm layer where airplanes fly
- contains the ozone laye that protects us from th Sun's harmful rays

TROPOSPHERE
- up to 12 miles (20 km) above the Earth
- weather layer, where rain clouds form

E OXYGEN CYCLE

'ast amount of oxygen exists
oceans, rocks, and the
iosphere. Oxygen created
plant photosynthesis
ances oxygen used up by
nbustion or animals
athing.

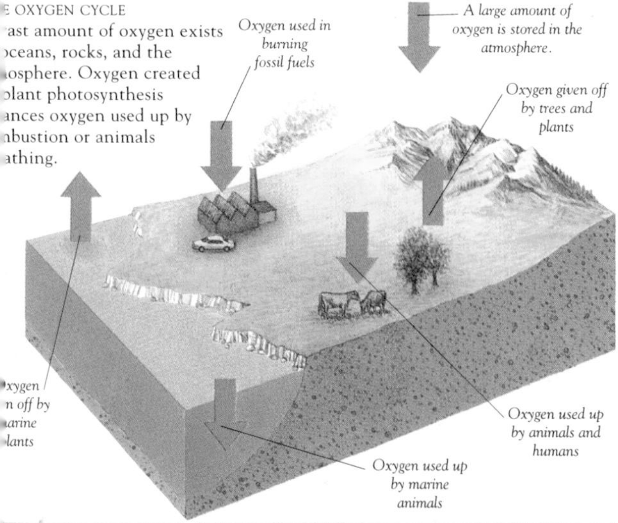

TMOSPHERE FACTS

The troposphere is
e densest layer of the
mosphere.

Ozone is a type of
ygen that absorbs
maging ultraviolet
ys from the Sun.

Humans can live and
eathe normally only
the troposphere layer.

COMPOSITION OF THE LOWER ATMOSPHERE

Although nitrogen makes up most of the air we breathe, oxygen is the essential gas for all animal and human life. Nitrogen is simply breathed in and out. Other gases, such as argon and carbon dioxide, make up less than 1 percent.

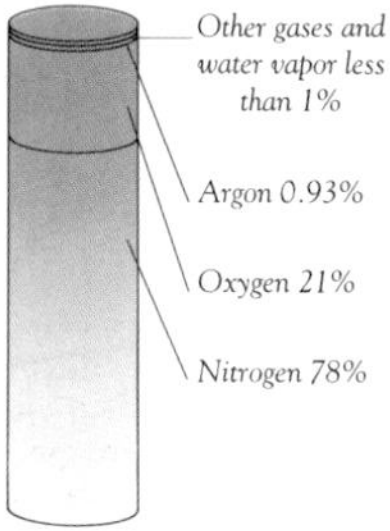

MAPPING THE EARTH

MAPS HELP US SEE what the Earth looks like. A map uses symbols to represent different features of the Earth. A technique called projection can transfer th curved surface of the globe onto a flat sheet of paper Aerial photographs help make maps that show valle and hills. On a larger sca satellite photograp help mapmake show how t Earth look from spac

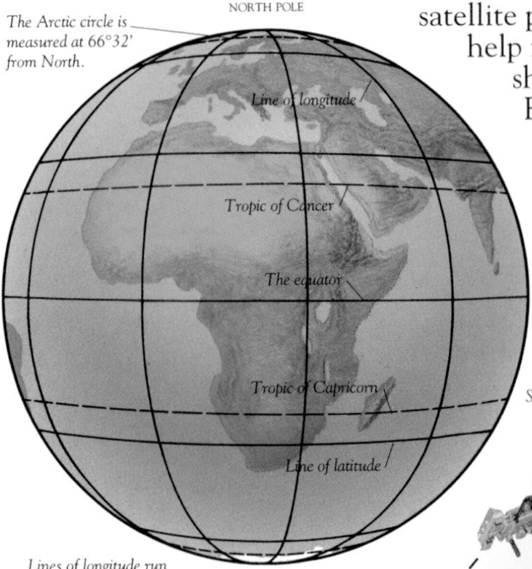

Lines of longitude run between the north and south poles, and lines of latitude are parallel to the equator.

EARTH'S GRI
Imaginary l
of longitud
and latitu
form a gri
the Earth.
grid is used
aid navigati

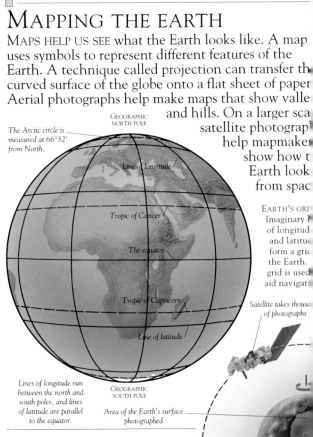

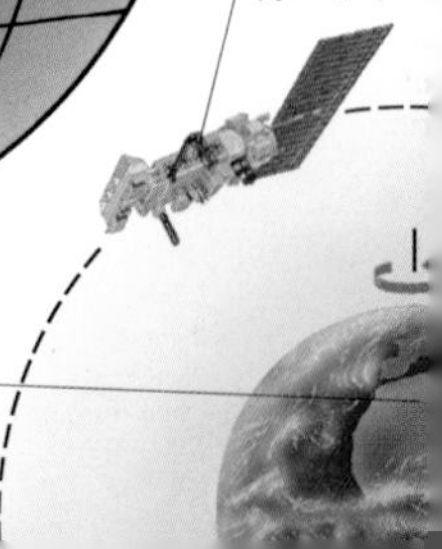

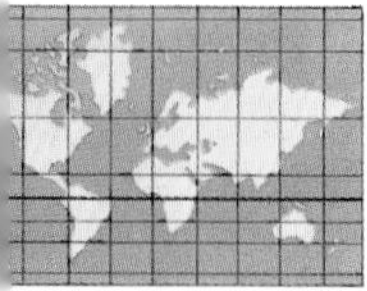

MERCATOR'S PROJECTION

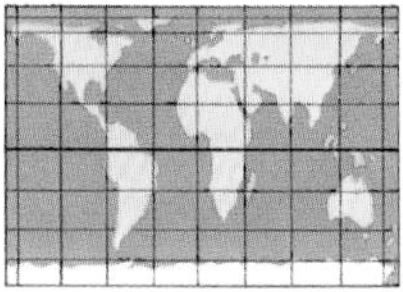

PETERS' PROJECTION

MAP PROJECTIONS
Mercator's map of 1569 distorted the land area of the continents – Greenland appeared larger than Africa. Peters' map shows the right land area but the shape of the continents is incorrect.

E WORLD'S CONTINENTS
e Earth is divided
ɔ seven land
sses or
tinents.

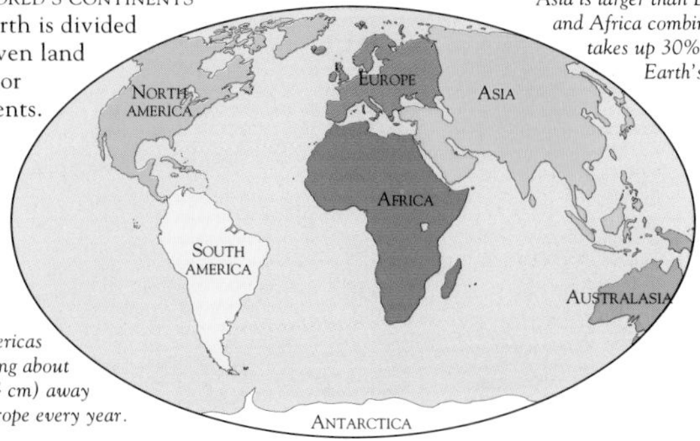

Asia is larger than Europe and Africa combined. It takes up 30% of the Earth's land.

Americas moving about in (4 cm) away ı Europe every year.

TELLITE MAPPING
ıile orbiting the Earth, ellites photograph the net in sections. The arate images are nbined to give a clear ture of the Earth.

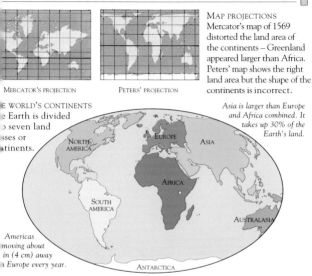

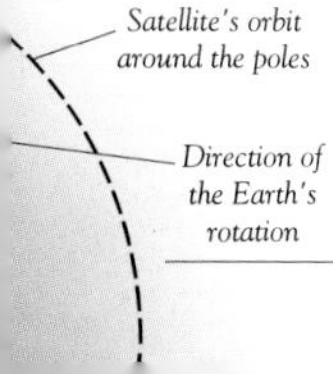

THE SIZE OF THE CONTINENTS		
CONTINENT	AREA IN SQ KM	AREA IN SQ MILES
Asia	44,485,900	17,176,100
Africa	30, 269,680	11,687,180
North America	24,235,280	9,357,290
South America	17,820,770	6,880,630
Antarctica	13,209,000	5,100,020
Europe	10,530,750	4,065,940
Australasia	8,924,100	3,445,610

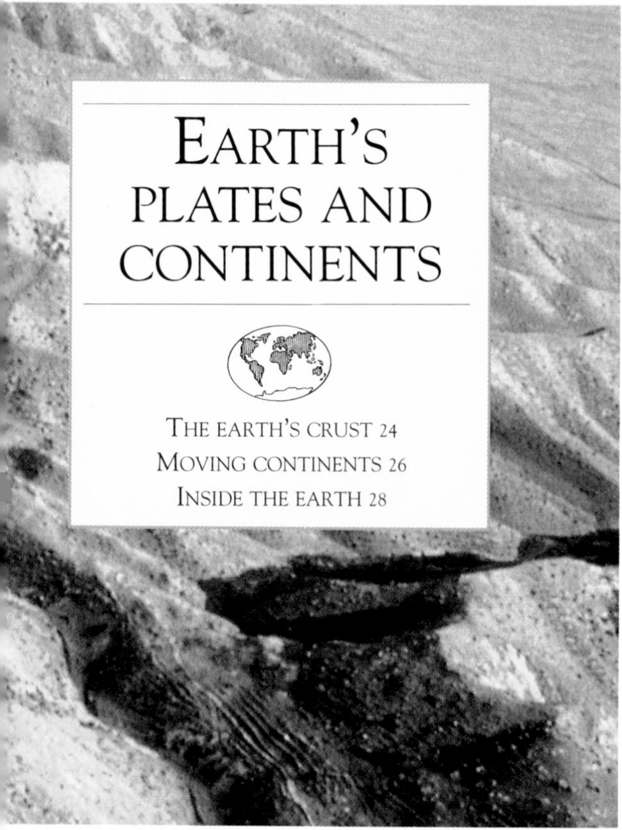

EARTH'S PLATES AND CONTINENTS

THE EARTH'S CRUST

EARTH'S SURFACE IS covered by a thin layer of rock called crust. Rocky crust standing above sea level forms islands and continents. The lithosphere is in pieces, or plates, that move slowly all the time. Whe two plates meet they may slide past each other or one may go under another. New crust forms at ocean ridges and old crust melts into the mantle.

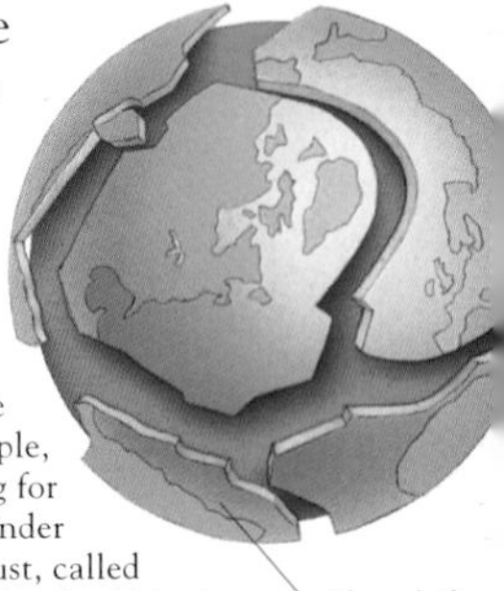

The rock plate the Earth's cru together like pi of a jigsaw puz

PLATE FACTS

- Earth's plates "float" on a layer of mantle called asthenosphere.
- The size of the Earth doesn't change – new crust produced equals older crust consumed.

EARTH'S SKIN

Earth's crust, like the skin of an apple, is a thin covering for what is inside. Under the ocean the crust, called oceanic crust, is 4 miles (6 km) thick, but under mountain ranges the continental crust can be 40 miles (64 km) thick.

CROSS-SECTION THROUGH THE EARTH'S CRUST

This section through the Earth's crust at the Equator shows the landscape and the direction of plate movement at plate boundaries.

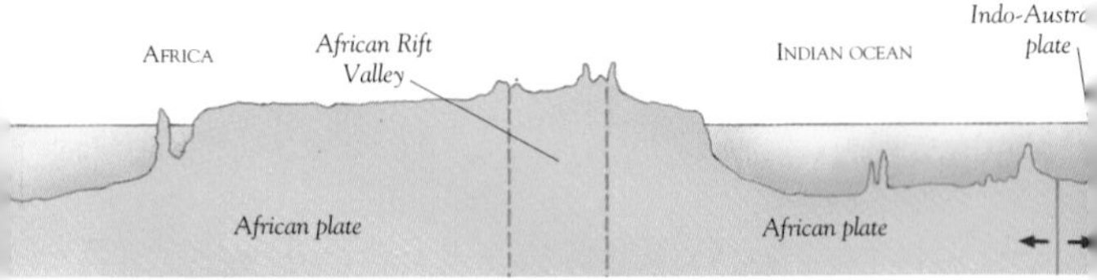

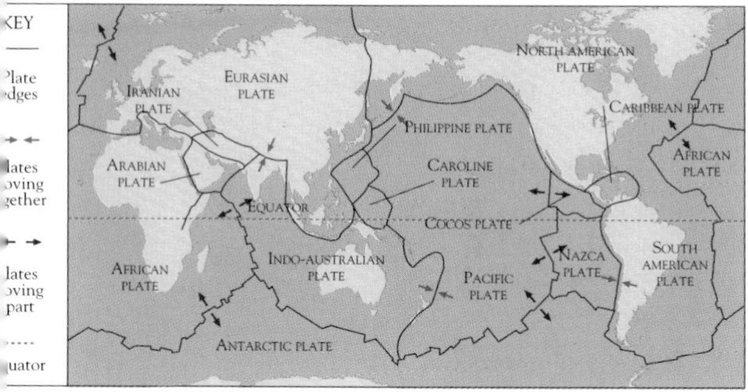

KEY

Plate
edges

Plates
moving
together

Plates
moving
apart

Equator

TES OF THE WORLD

e surface of the Earth
15 large plates. A
e can include both
tinental lithosphere
oceanic lithosphere.
as such as Australia
n the middle of a
e, while others, like
and, have a plate
ndary through them.

MOVEMENT OF THE EARTH'S PLATES			
PLATE NAMES	DIRECTION OF MOVEMENT	RATE OF MOVEMENT	
		CM PER YEAR	IN PER YEAR
Pacific/Nazca	apart	18.3	7.2
Cocos/Pacific	apart	11.7	4.6
Nazca/South American	together	11.2	4.4
Pacific/Indo-Australian	together	10.5	4.1
Pacific/Antarctic	apart	10.3	4.0

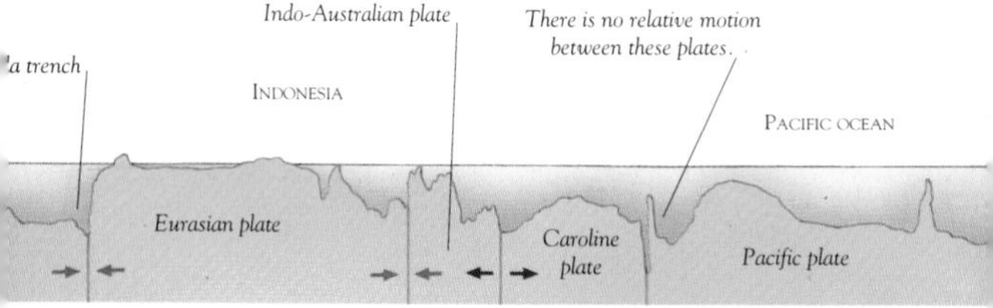

MOVING CONTINENTS

EARTH'S CONTINENTS can be rearranged to fit together like pieces of a jigsaw puzzle. This idea made scientists think that they once formed a giant landmass, Pangaea. This "supercontinent" broke up and the continents drifted, over millions of years, to where they are now. This is continental drift, or plate tectonics, theory. Continents move as the Earth's plates move, sliding along on the asthenosphere, a layer of soft mantle.

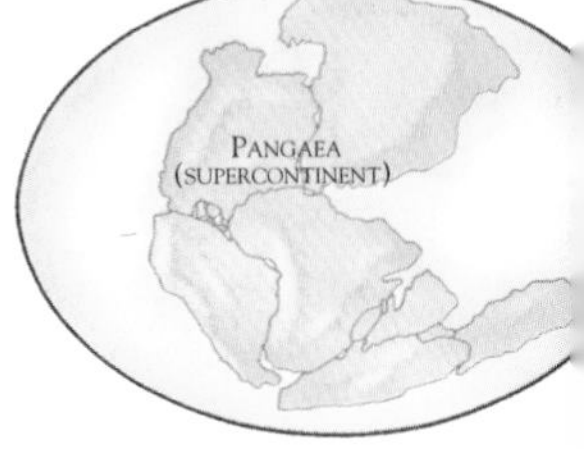

250 MILLION YEARS AGO

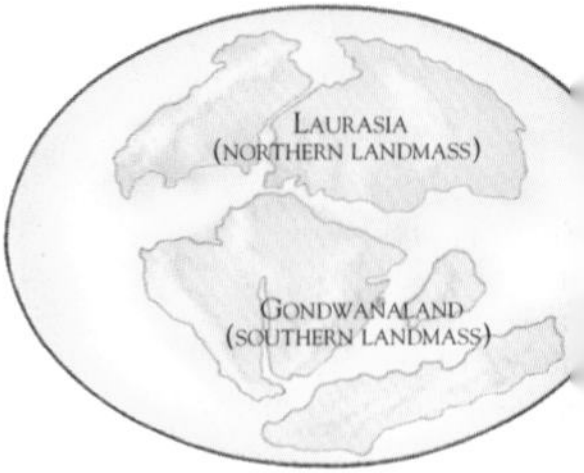

120 MILLION YEARS AGO

CONTINENTAL DRIFT
When Pangaea broke up, new continents emerged. The outlines South America and Africa appear

CROSS-SECTION THROUGH THE EARTH'S CRUST

PACIFIC OCEAN

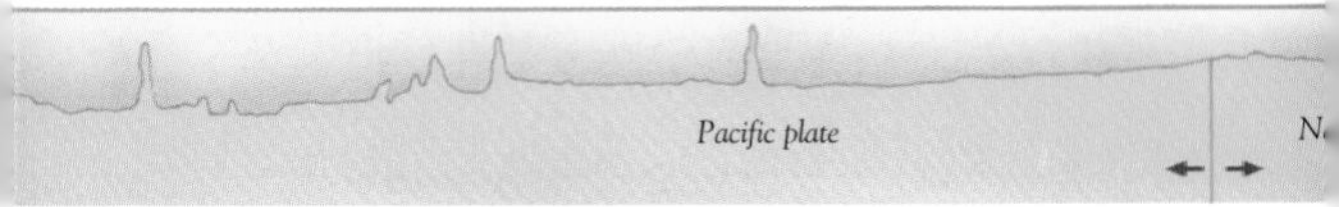

LATE BOUNDARIES

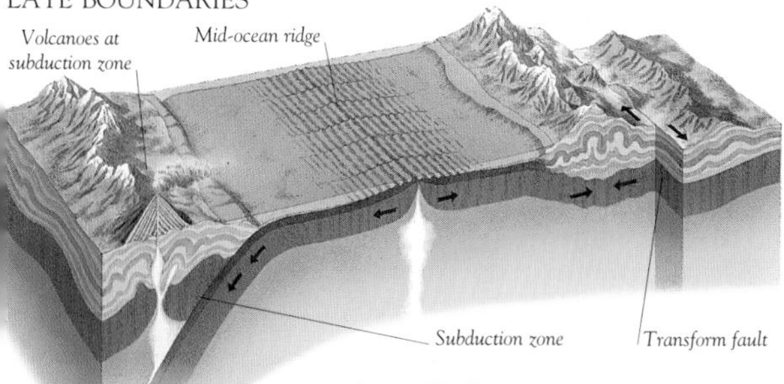

'HERE PLATES MEET
t a transform fault, ates slide past one ıother. A subduction ne is where two plates ·llide. One plate is rced into the mantle d molten rock material es to form volcanoes. : a mid-ocean ridge new ıst rises between plates.

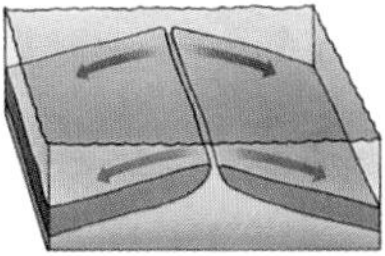

DRAG THEORY
Heat inside the Earth forces the mantle to rise in convection cells. It erupts at mid-ocean ridges, dragging plates apart.

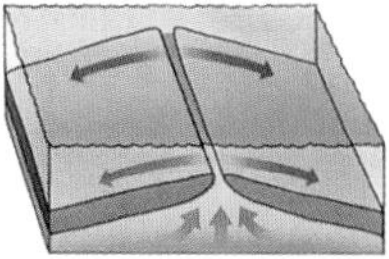

PULL THEORY
Rising molten rock cools and solidifies. This denser rock sinks at trenches and gravity pulls the plate down.

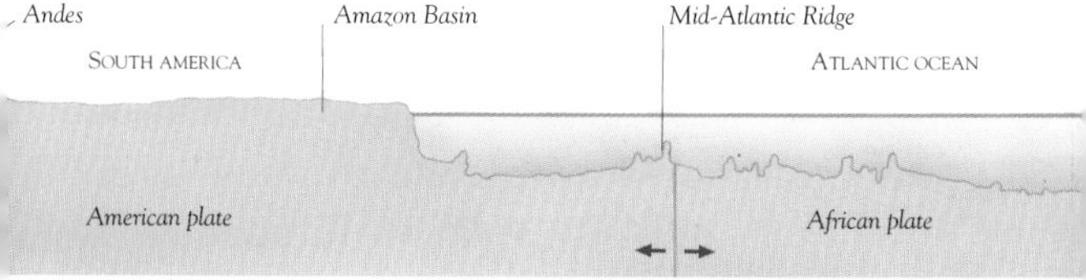

INSIDE THE EARTH

THE INTERIOR OF the Earth has four major layers. O the outside is the crust made of familiar soil and roc Under this is the mantle, which is solid ro with a molten layer at the to The inside or core of th Earth has two sections: an ou core of thick fluid, and a so inner core.

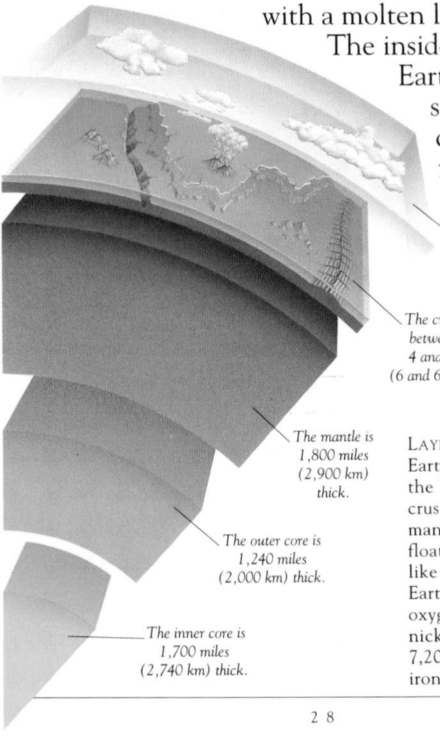

The atmosphere stretc about 400 miles (640 km) into spac
The crust varies between about 4 and 40 miles (6 and 64 km) thick.
The mantle is 1,800 miles (2,900 km) thick.
The outer core is 1,240 miles (2,000 km) thick.
The inner core is 1,700 miles (2,740 km) thick.

LAYERS OF THE EARTH
Earth's outer shell is call the lithosphere. This is t crust and part of the upp mantle. The lithosphere floats on the asthenosph like an iceberg on the se Earth's outer core is mos oxygen, liquid iron, and nickel. Its inner core, ab 7,200°F (4,000°C), is sol iron and nickel.

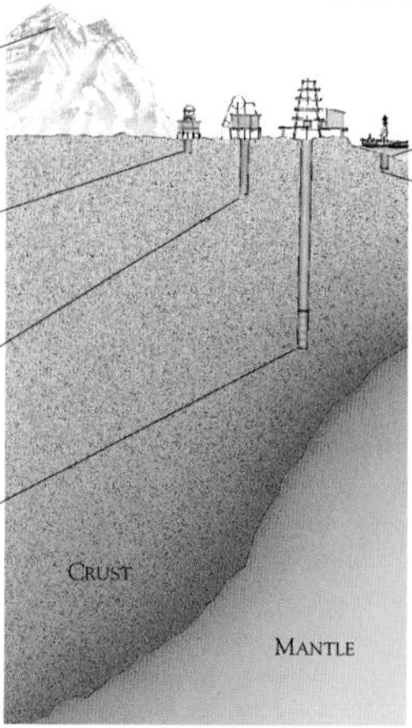
Aount Everest
reaches up
miles (8.85 km).
deep coal mine
ıes down less than
52 miles (1 km)
elow sea level.
ıe deepest mine
ches down 2.34
ıiles (3.8 km).
he deepest hole
has reached
miles (12 km).
CRUST
MANTLE
Sea level
The deepest ocean
drilling has reached
1.05 miles (1.7 km).

UNDER THE CRUST
Clues to the interior of the Earth come from boreholes. These have reached only as far as the crust. They reveal a dense lower layer and a less dense upper layer to the crust. There is also evidence of a boundary between crust and mantle called the Moho.

.UST FACTS

f an excavator could : a hole through the rth at 39 in (1 m) r minute, it would e 24 years to reach : other side.

Western Deep Gold ne in South Africa is : world's deepest ne. It is 2.36 miles 8 km) deep.

The Earth's crust is inly granitelike rock.

COMPOSTION OF THE EARTH'S CRUST
Light elements such as silicon, oxygen, and aluminum make up the Earth's crust. Oceanic crust is mostly basalt (which also contains magnesium and iron). Continental crust is composed of granitelike rocks. These may have formed from recycled basaltic ocean crust.

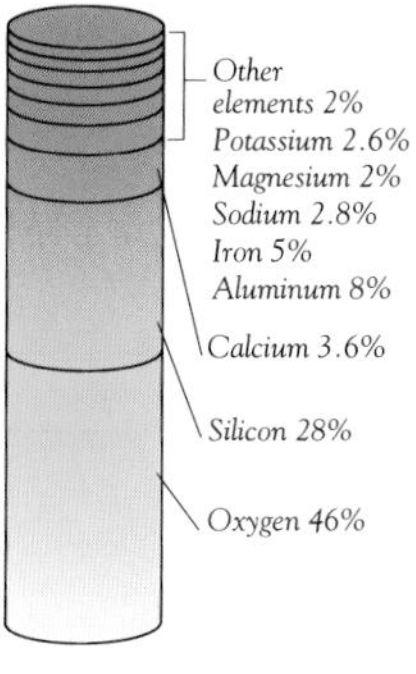
Other
elements 2%
Potassium 2.6%
Magnesium 2%
Sodium 2.8%
Iron 5%
Aluminum 8%
Calcium 3.6%
Silicon 28%
Oxygen 46%

VOLCANOES

EXTINCT VOLCANO
Castle Rock, Edinburgh is an extinct volcano. It has not erupted for 340 million years. An extinct volcano such as this is not expected to erupt again.

DORMANT VOLCANO
If scientists believe a volcano may erupt again, perhaps because it gives off volcanic gases, it is called dormant. Mt. Rainier, WA, is considered dormant.

THE EARTH'S VOLCANOES

MOST VOLCANOES are found at plate boundarie near the Pacific coast or mid-ocean ridges. Here fractures in the lithosphe allow molten rock, callec magma, to rise from the mantle inside the Earth. Magma is known as lava when it flows out of a volcano. Ash, steam, anc gas also spew out from a volcano and can cause a great deal of destruction.

COMPARING ERUPTIONS
One way to compare the size of different volcanic eruptions is to measure the amount of ash thrown out during an eruption.

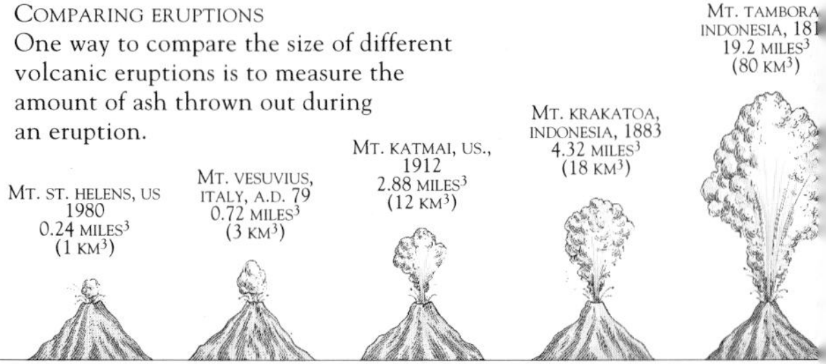

.CANO SITES

eneral, volcanoes, like hquakes, occur near e boundaries. canoes are ned by plate ruction or presence of t spot erneath plate.

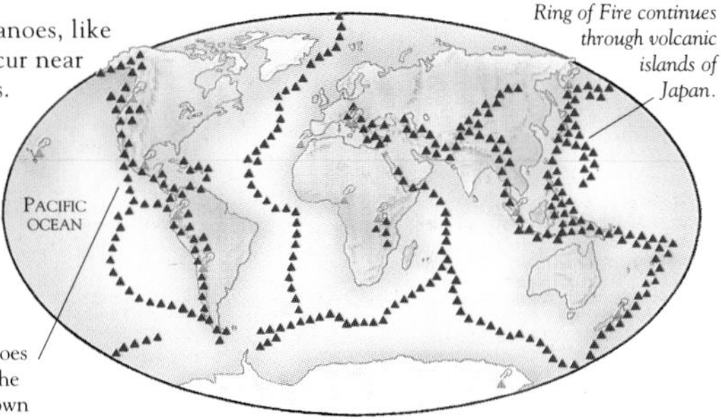

Ring of Fire continues through volcanic islands of Japan.

hain of volcanoes d the edge of the c Ocean is known "Ring of Fire."

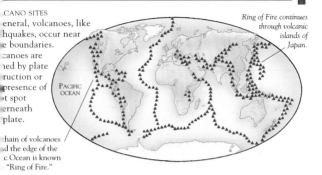

PEII

.D 79 Mt. Vesuvius erupted, ing the town of Pompeii er dust and ash. The two-day tion killed 2,000 people with onous gases and hot ash.

LARGEST VOLCANIC EXPLOSIONS

The Volcanic Explosivity Index (V.E.I.) grades eruptions from 0 to 8. The scale is based on the height of the dust cloud, the volume of tephra (debris ejected by a volcano), and an account of the severity of the eruption. Any eruption above 5 on the scale is very large and violent. So far, there has never been an eruption of 8.

VOLCANO	DATE	V.E.I.
Crater Lake, Oregon	c.4895 B.C.	7
Towada, Honshu, Japan	915	5
Oraefajokull, Iceland	1362	6
Tambora, Indonesia	1815	7
Krakatoa, Indonesia	1883	6
Santa Maria, Guatemala	1902	6
Katmai, US	1912	6
Mt. St. Helens, Washington	1980	5

VOLCANO SHAPES

NOT ALL VOLCANOES are the same. Some are cone-shaped and others are almost flat. The shape of the volcano depends on the kind of lava that erupts. Soupy, nonviscous lava spreads quickly before hardening, but stiff, viscous lav piles up near the volcanic vent. Volcanoes usually appear near plate boundaries, but they also form at hot spots such as in Hawaii. Volcanoes also exist under the ocean at plate edges.

ICELAND'S RIFT
Skaftar fissure in Iceland lies where two plates are moving apart. It is part of a 16-mile (27-km) rift along the plates' edges.

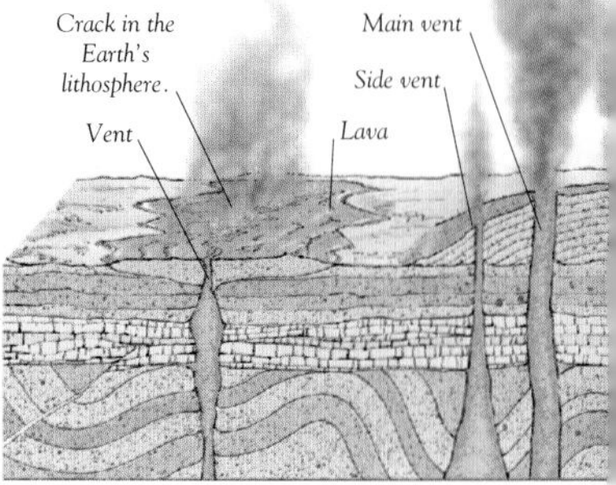

FISSURE VOLCANO
This type of volcano arises from a long crack in the lithosphere. Nonviscous lava flows out to form a plateau.

SHIELD VOLCANO
A large, gently slop volcanic cone. It g when nonviscous l erupts from a centr vent or side vents.

VOLCANO FACTS

- Kilauea, Hawaii is one of the world's most active volcanoes.
- There are about 1,300 active volcanoes in the world.

ORST VOLCANIC ERUPTIONS		
VOLCANO	DATE	HUMAN DEATHS
ıbora	1815	92,000
Pelée	1902	40,000
ĸatoa	1883	36,000
ado del Ruiz	1985	23,000

COMPOSITE VOLCANO

Cone-shaped volcanoes build up from viscous lava. Inside are layers of thick lava and ash from previous eruptions. Gas pressure inside the volcano magma chamber causes violent eruptions.

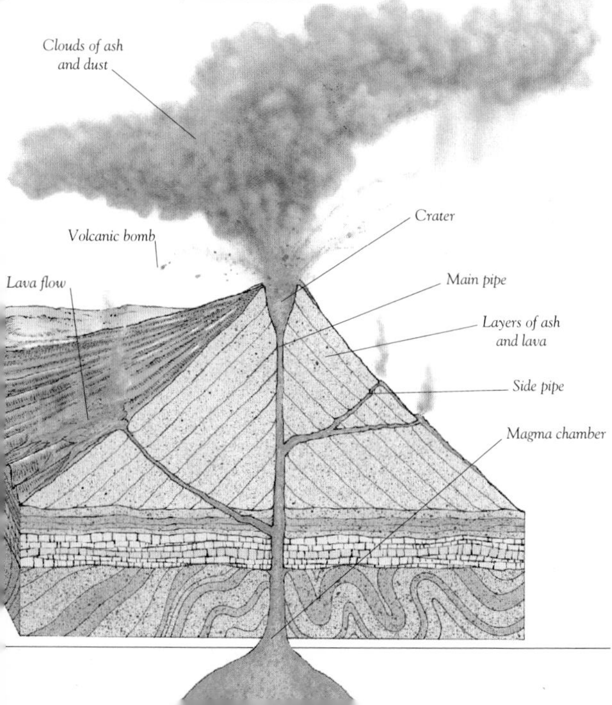

EXPLOSIVE VOLCANOES

SOMETIMES VOLCANOES explode violently. Volcanoe that form from viscous lava are most likely to do thi Viscous lava tends to plug up volcanic vents. When pressure in the magma chamber increases, th lava is blown out. Pieces of rock and a great deal of ash are hurled high into the air. Clouds of ash ar pumice flow like hot avalanches down the sides of the volcano. Mudflows (also called lahars) are mixture of water and ash. They travel at great speed and engulf everything in their paths.

NUÉE ARDENTE
An explosive eruption can cause a glowing ash cloud or *nuée ardente*.

BEFORE MT. ST. HELENS ERUPTED

MT. ST. HELENS
In the Cascades a peaceful-looking volcano erupted after 123 years of dormancy. A warning came when one side of the volcano began to bulge as the magma rose. A gas explosion lasting 9 hours and a landslide ensued. An ash cloud over 580 miles^2 (1,500 km^2) caused darkness. Melted snow and ash made mudflows.

ERUPTION ON MAY 18TH 198

VASTATING MUDFLOWS
ıen Ruiz volcano in Colombia erupted in 35, the snow melted around its summit. A xture of water, dust, and ash fast turned to d and buried the nearby city of Armero. More n 22,000 people were drowned in the mud.

PUMICE FACTS

- Pumice is really light-weight frothed glass that is able to float on water.
- Pumice is used in industry as an abrasive for soft metals. It is also used for insulation in some buildings.

RODUCTS OF EXPLOSIVE VOLCANOES

ASH
ava particles arger than dust over the land.

LAPILLI
Lava ejected in pea-sized pieces is called lapilli.

UMICE
umice is light-veight lava illed with holes.

BOMB
Bomb-shaped lava forms as it flies in the air.

PELE'S HAIR
Sometimes drops of liquid lava blow into fine spiky strands. The threads form needles of volcanic glass. They are named Pele's hair after Pele, the Hawaiian goddess of volcanoes.

NONEXPLOSIVE VOLCANOES

SOME VOLCANOES arise from fissures. Nonviscous lava flows for long distances before cooling. It builds broad plateaus or low-sided volcanoes. This kind of volcano forms at plate edges, mostly under the ocean. A hot spot volcano bursts through the middle of a plate; it is not related to plate margins.

SPREADING RIDGES
Flows of basalt lava from fissures form mountains along the edges separating plates. These spreadin ridges are usually underwater. In some places, such as Iceland, lav erupts along the crack forming ri mountains above sea level.

TYPES OF LAVA

PAHOEHOE LAVA
Lava with a wrinkled skin is called pahoehoe. This nonviscous lava cools to form a "ropy" surface. Such flows of basalt pahoehoe lava are common in Hawaii.

AA LAVA
This is an Hawaiian word for slow-moving, viscous lava. When aa lava solidifies it has a rough, jagged surface that is also described as blocky.

.WAIIAN VOLCANOES
places called hot spots, notably Hawaii, magma rises to create : fountains and fire curtains.

BASALT COLUMNS
Northern Ireland's Giant's Causeway is made of mostly hexagonal columns of basalt rock. They formed as thick lava flows cooled and vertical shrinkage cracks developed.

)T SPOT VOLCANOES
e Earth's plates move slowly over t spots in the crust. Magma rises, ıching through the lithosphere to m a new island. In Hawaii, hot ›ts have built a chain of islands.

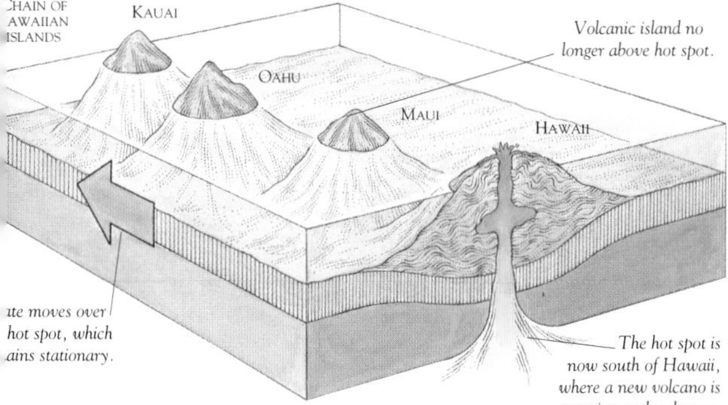
KAUAI
OAHU
MAUI
HAWAII
Volcanic island no longer above hot spot.
ıte moves over hot spot, which ains stationary.
The hot spot is now south of Hawaii, where a new volcano is growing under the sea.

CHAIN OF AWAIIAN ISLANDS

OLD FAITHFUL
This geyser is in Yellowstone Park, Wyoming. It has shot out a column of boiling water and steam every hour for the last 100 years.

VOLCANIC LANDSCAPES

MOVEMENT IN ROCKS undergroun can cause changes to the landsca above. The combination of heat and water in the Earth's rocks produces various phenomena. Molten rock erupting out of the Earth brings gases, mineral deposits, and water with it. Mud pools, hot springs, and geysers fo when the gases and water escape. Minerals dissolved in the hot waters precipitate to form cone-like or terraced deposits of rocks.

THE LANDSCAPE AROUND VOLCANOES

Steaming hot water

HOT SPRINGS
Magma warms water in cracks in the rock. Water returns to the surface as a hot spring.

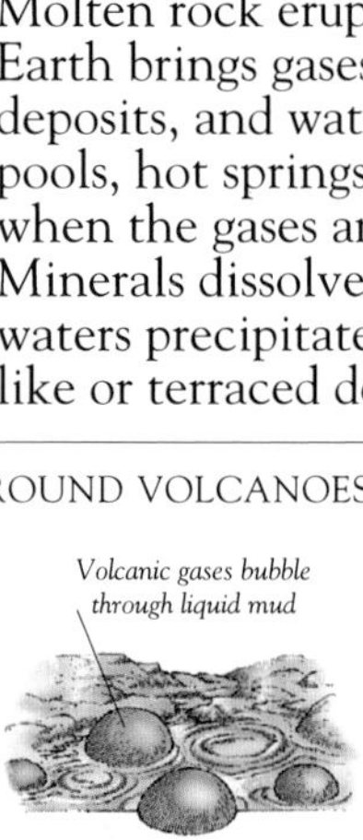
Volcanic gases bubble through liquid mud

MUD POOLS
Steam, particles of rock, and volcanic gases bubble through pools of liquid mud.

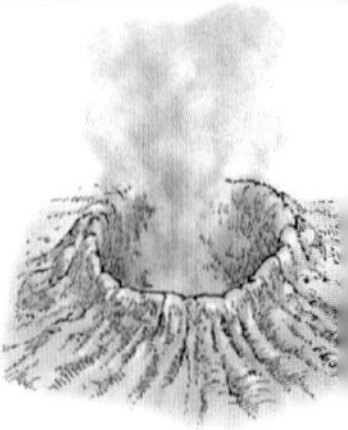

FUMAROLES
Vents, or fumeroles, allow steam and oth gases to escape from cooling rocks.

PILLOW LAVA
When lava erupts underwater it can produce these rounded shapes, which are known as pillow lava. The seawater cools the lava rapidly, so that as it solidifies a crust forms around each lump. The lava formed is typically an igneous rock called basalt.

A NEW ISLAND
In 1963, off the southern coast of Iceland, a new volcanic island rose from the ocean floor. This island, Surtsey, formed from the buildup of lava flows. Seawater interacting with lava produced explosions and huge amounts of steam.

'SERS
ıt from magma cham- ; causes ground water ›oil, erupting as jets of m and water.

TERRACES
Minerals, dissolved in heated ground water, are deposited in layers that rise around a vent.

GEYSER FACTS

- The tallest geyser is Yellowstone National Park's Steamboat Geyser in Wyoming. It reaches 195–380 ft (60–115 m).
- Strokkur Geyser in Iceland spurts every 10 to 15 minutes.
- In 1904, Waimangu Geyser, New Zealand erupted to a height of 1,500 ft (460 m).

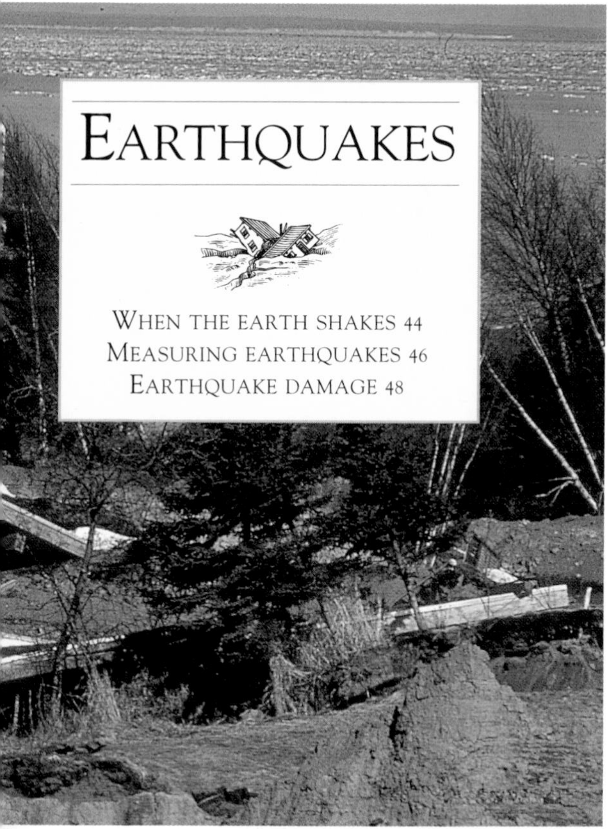

EARTHQUAKES

WHEN THE EARTH SHAKES

MORE THAN A million times a year, the Earth's crust suddenly shakes during an earthquake. Most of the world's earthquakes are fairly slight. A mild earthquake can feel like a truck passing; a severe one can destroy roads and buildings and cause the sea to rise in huge waves. Earthquakes often happen near volcanoes and young mountain ranges: at the edges of the Earth's plates.

A SEVERE EARTHQUAKE
The city of San Francisco was shaker
a devastating earthquake in 1906. O
the chimney stacks were left stand

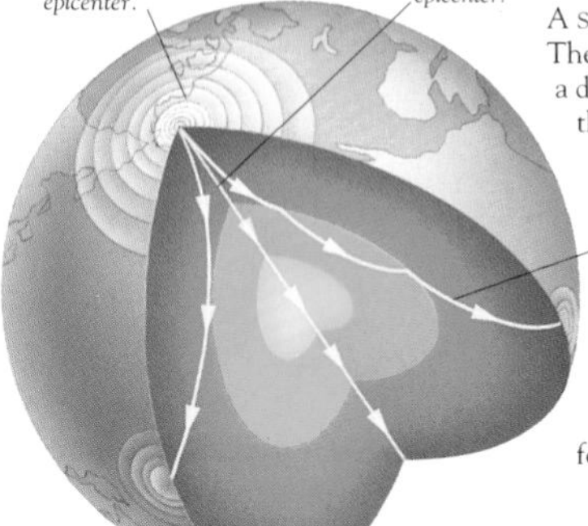
Surface waves radiate from the epicenter.
The focus is below the epicenter.
Shock waves go through the Earth and up to the surface.

CENTER OF AN EARTHQU
The earthquake is stronge
the focus. At the epicenter,
point on the surface above
focus, the crust shakes and sends
shock waves. Internal waves b
as they travel through the Ea

.THQUAKE FAULT ZONES

thquakes occur at cracks in the osphere called faults. Deep earthquakes place where one e is sliding er ther.

Many earthquakes occur on the northeast coast of Asia. This is at the boundary of two of the Earth's plates.

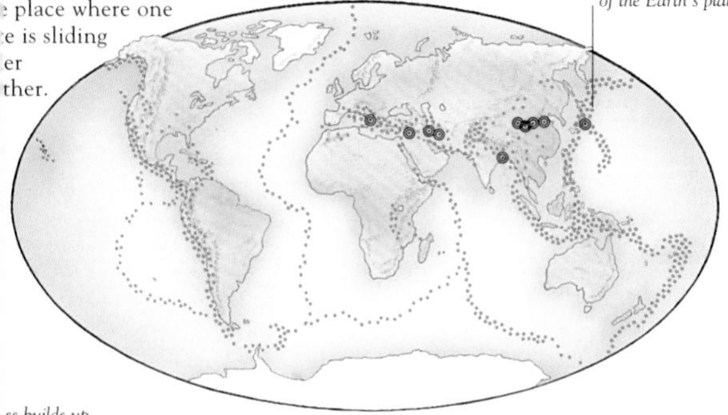

ss builds up rocks along fault line.

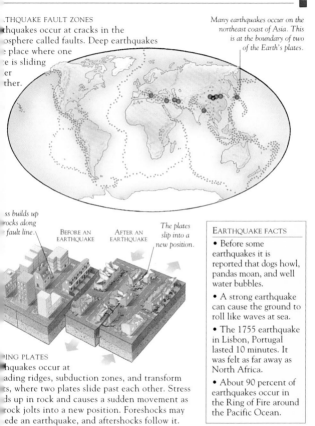

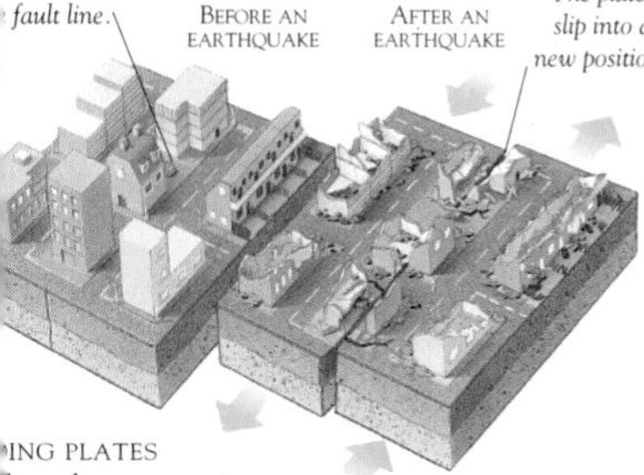

The plates slip into a new position.

ING PLATES

hquakes occur at ading ridges, subduction zones, and transform ts, where two plates slide past each other. Stress ds up in rock and causes a sudden movement as rock jolts into a new position. Foreshocks may ede an earthquake, and aftershocks follow it.

EARTHQUAKE FACTS

- Before some earthquakes it is reported that dogs howl, pandas moan, and well water bubbles.
- A strong earthquake can cause the ground to roll like waves at sea.
- The 1755 earthquake in Lisbon, Portugal lasted 10 minutes. It was felt as far away as North Africa.
- About 90 percent of earthquakes occur in the Ring of Fire around the Pacific Ocean.

EARTHQUAKE DESTRUCTION
In 1994 in the city of Los Angeles an earthquake caused devastation. Roads and buildings collapsed, water mains and gas pipes burst, and fires began in the city. Many buildings in Los Angeles were built to be earthquake-proof and so did not suffer very much damage. The earthquake measured 5.7 on the Richter scale.

MEASURING EARTHQUAKES

SCIENTISTS WHO STUDY earthquakes are known as seismologists (*seismos* is the Greek word for earthquakes Seismologists monitor the vibrations or shock waves that pass through the Earth using an instrument called a seismometer. Predicting earthquakes is very difficult Scientists look for warnings such as bulges or small crac on the surface of the groun

MERCALLI SCALE
Giuseppe Mercalli (1850–1914) devised a method of grading earthquakes based on the observation of their effects. Using this scale enables the amount of shaking, or intensity, of different earthquakes to be easily compared. On Mercalli's scale earthquakes are graded from 1 to 12.

1 • detected by instruments
2 • felt by people resting
3 • hanging lamps sway
4 • felt by people indoors
• plates, windows rattle
• parked cars rock

5 • buildings tremble
• felt by most peopl
• liquids spill
6 • movement felt by
• pictures fall off wa
• windows break

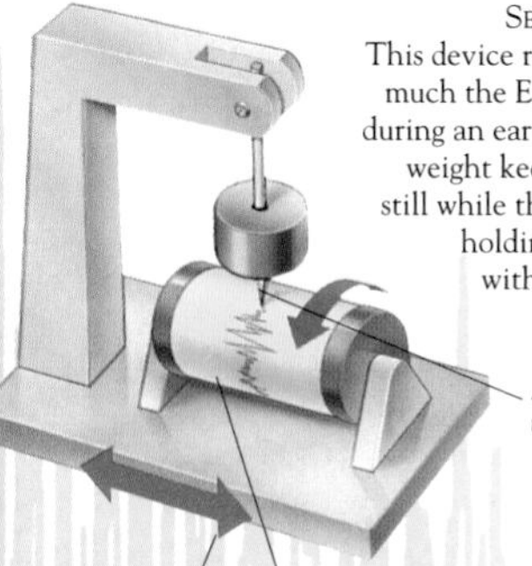

SEISMOMETER
This device records how much the Earth shakes during an earthquake. A weight keeps the pen still while the machine holding it moves with the Earth.

A pen records the movement on a rotating drum.

[Ba]se moves with the [h]orizontal motion of the Earth.

The recording from a seismometer is called a seismogram. It shows an amplified wave form, resulting from the motion of the Earth's surface.

RICHTER SCALE

The amount of energy released by an earthquake can be measured on the Richter scale. An increase of 1.0 on the scale represents a tenfold increase in energy.

EARTHQUAKE	DATE	RICHTER SCALE
North Peru	1970	7.7
Mexico City	1985	7.8
Erzincan	1939	7.9
Tangshan	1976	8.0
Tokyo	1923	8.3
Kansu	1920	8.6

- bricks and tiles fall
- chimneys crack
- difficult to stand
- steering cars difficult
- tree branches snap
- chimneys fall

9
- some buildings collapse
- ground cracks
- mud oozes from ground

10
- underground pipes burst
- river water spills out
- most buildings collapse

11
- bridges collapse
- railroad tracks buckle
- landslides occur

12
- near total destruction
- rivers change course
- waves seen on ground

EARTHQUAKE DAMAGE

IN GENERAL, great loss of life during an earthquake can be avoided. It is often not the Earth's shaking that kills people but falling buildings, particularly poorly constructed ones. Landslides and tsunamis also cause a lot of damage. During an average earthquake, it is best to stay indoors in a doorway or under a sturdy table. Falling masonry is a hazard outdoors.

ESTIMATED LIVES LOST AS A RESULT OF RECENT EARTHQUAKES

PLACE	YEAR	ESTIMATED DEATHS
Tangshan, China	1976	695,000
Kansu, China	1920	100,000
Tokyo, Japan	1976	99,000
Messina, Italy	1908	80,000
Armenia	1988	55,000
Northwest Iran	1990	40,000
Erzincan, Turkey	1939	30,000

FIRE HAZARD
Fire poses a great danger followin an earthquake. Gas leaks and oil spills can lead to large fires like t in San Francisco in 1989.

JAPANESE PRINT SHOWING A TSUNAMI TALLER THAN MT. FUJI IN JAPAN.

TSUNAMIS

An earthquake on the continental shelf can start a wave at sea. Such waves have low height in deep water. As the wave nears shore its front slows and water behind builds up to form a huge tsunami.

TSUNAMI FACTS

- The highest tsunami wave was 279 ft (85 m) high. It struck Ishigaki Island, Japan in 1971.
- In the open ocean, a tsunami can travel at speeds of 370 miles (600 km) per hour.

DSLIDES

se rock and debris may be odged by an earthquake cause landslides as in ska in 1964. Avalanches may be triggered by the nd shaking. Mudflows or rs can result from rain or snow mixing with loosened soil.

EARTHQUAKE-PROOF BUILDINGS

A great deal of damage is caused by buildings collapsing during earthquakes. In earthquake-prone San Francisco and Japan there are safety guidelines that all new buildings must meet. Wooden buildings are replaced with concrete and concrete and steel foundations are used.

Pyramid-shaped buildings are built to withstand stress.

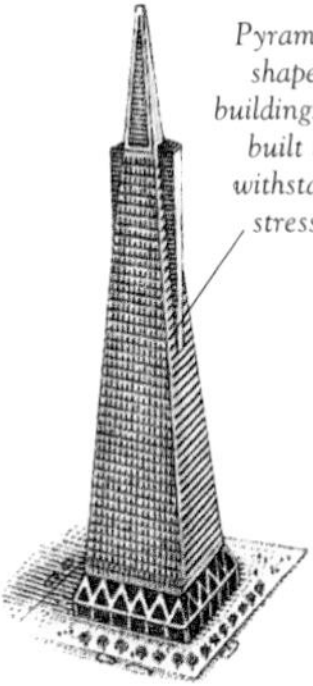

TRANSAMERICA BUILDING, SAN FRANCISCO

The central column of the pagoda absorbs the shaking.

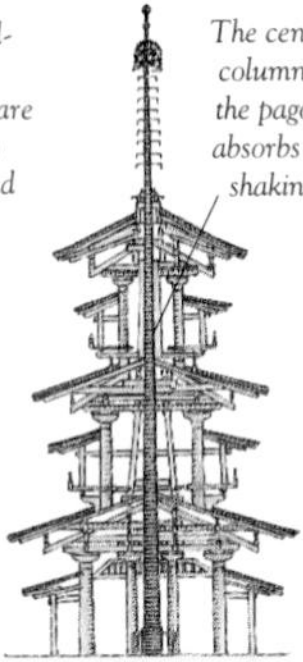

ANCIENT BUDDHIST PAGODA, JAPAN

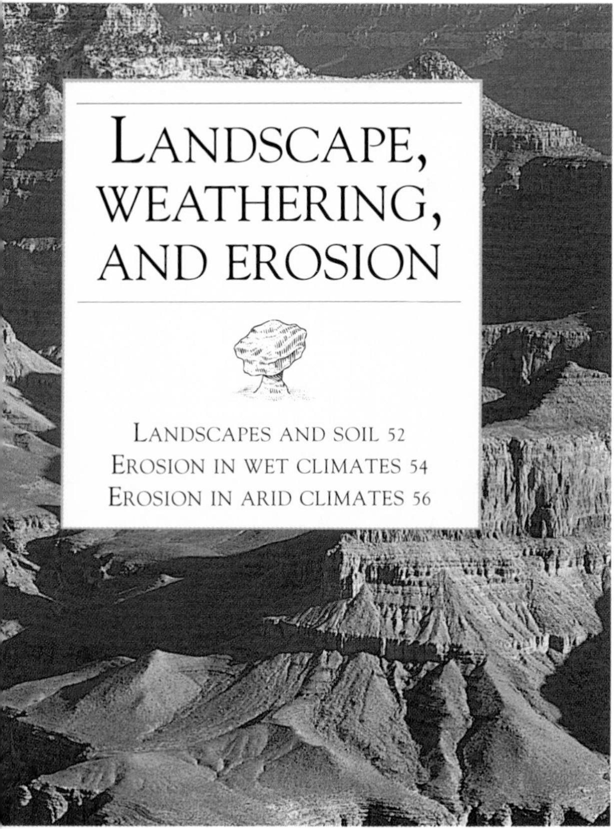

LANDSCAPE, WEATHERING, AND EROSION

LANDSCAPES AND SOIL

WEATHERING PROCESSES are primarily responsible fo soil development. These processes take thousands of years. Climate, vegetation, and rock type determine what type of soil forms. Soil contains organic matter froı decaying plants and anim (humus) as well as sand, s and clay. It covers the landscape and provides a medium for plants to grow i

PEAT LANDSCAPE
This landscape is green and low-lying. Spongy peat soil is rich in humus from decayed bog plants. It retains water and nutrients easily.

SANDY LANDSCAPE
In arid (dry) sandy landscapes there is little vegetation. The soil contains hardly any organic material. Winds blow away small particles, leaving sand and stones.

TYPES OF SOIL
Chalky soil is thin and stony; water passes through it quickly. Water drains easily through sandy soil, washing out nutrients. Clay soil retains nutrients and moisture but is difficult for plants to take root in. Peat soil is acidic. It holds water and minerals.

CLAY

SAND

PEAT

CHALK

ɔIL FACTS

10.8 ft^3 (1 m^3) of soil
ay contain more than
billion animals.

Some soils in India,
frica, and Australia
'e 2 million years old.

It takes about 500
ːars for 1 in (2.5 cm)
ᶠ topsoil to form.

ıL PROFILE

lice of soil down to
bedrock is called a
ı profile. The profile
ws several layers, or
izons. The number
ɪ thickness of horizons
y with the soil type.

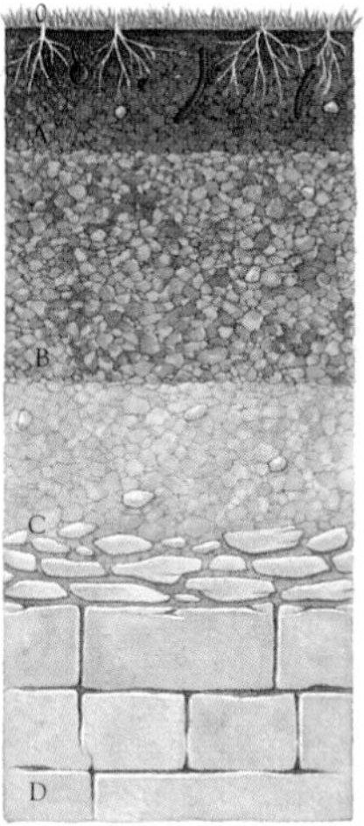

Horizon 0
- humus layer
- contains live and decaying plants and soil animals

Horizon A
- topsoil
- dark and fertile
- rich in humus

Horizon B
- subsoil
- contains minerals washed down from topsoil
- little organic matter
- lighter colored

Horizon C
- infertile layer
- composed of weathered parent rock

Horizon D
- bedrock (parent rock)
- source of soil's minerals

L CREEP AND EROSION

ıvity and water pull
. down a slope particle
particle. This is called
. creep. Plant roots
d soil and help to
vent it from wearing
ıy, or eroding. Over-
ːing and felling forests
h lead to soil erosion.

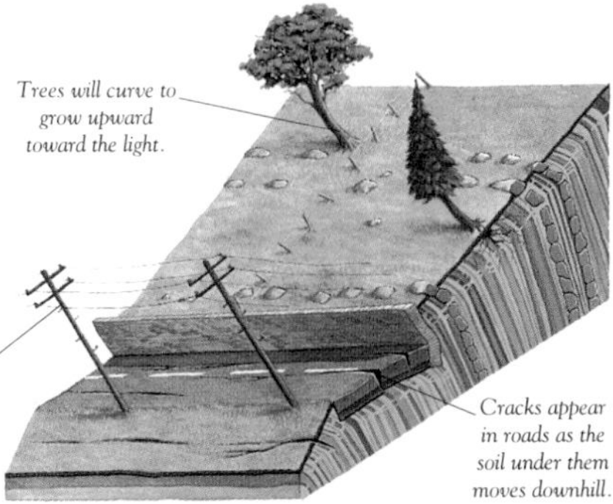

Trees will curve to grow upward toward the light.

Soil creep is indicated by leaning structures such as walls and telegraph poles.

Cracks appear in roads as the soil under them moves downhill.

EROSION IN WET CLIMATES

AS SOON AS ROCK is exposed on the Earth's surface, i is attacked by wind, water, or ice – a process known weathering. This prepares for erosion, when rock is broken down and removed. Weatherii can be either physical (wearin away the rock itself) or chemic (attacking the minerals in the rock). Climate and rock type determine the kind of weathering that occurs. In wet climates chemical weathering, mainly by rainwater, is domina

MOUNTAIN STREAM
Cascading over steep gradients, a swift-flowing stream wears away softer rocks. Harder rocks remain and create rocky outcrops. These become steep rapids or waterfalls.

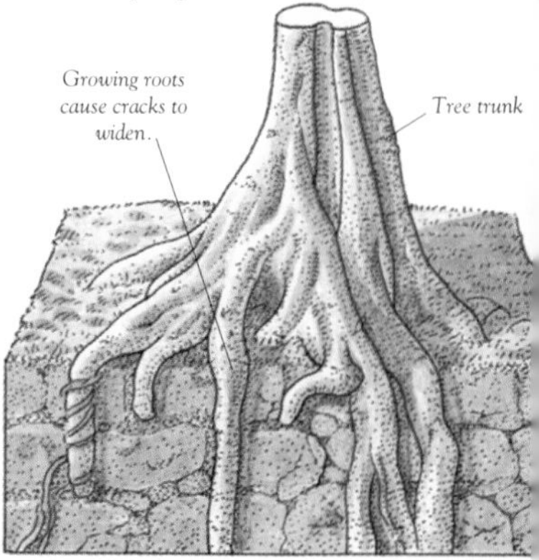

Growing roots cause cracks to widen.
Tree trunk

TREE-ROOT ACTION
As trees and other plants grow, their roots push down into small cracks in the rock. The cracks widen as the roots grow and eventually the rock breaks up.

)ST SHATTERING
is type of weathering
urs when water in
cks in the rock
ezes and expands.
its in the rock
len and the
k shatters.

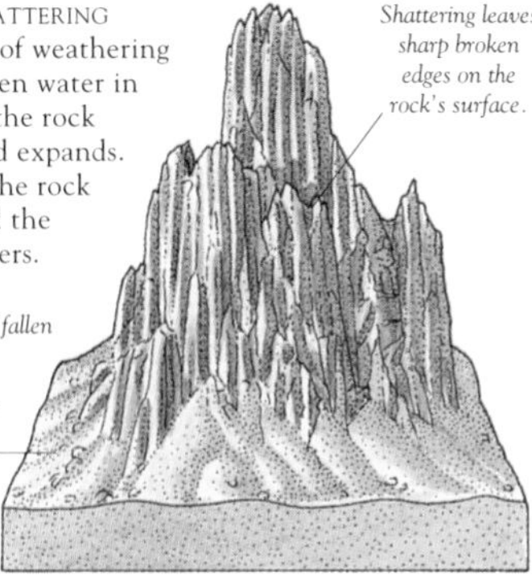

ERosion FACTS

- Acid rain can dissolve rock as deep as 98 ft (30 m) below the surface.
- Erosion is fastest in steep, rainy areas and semiarid areas with little vegetation.
- The rate of erosion for the whole of the world's land area is estimated to be 3.3 in (8.6 cm) every 1,000 years.

NTS AND GRIKES
d in rainwater seeps
) limestone joints and
olves the calcite in the
k. Ridges known as
ts and grooves known
rikes form in the rock.

ACID RAIN

Rainwater naturally contains a weak acid called carbonic acid. However, the burning of fossil fuels produces gases such as sulfur dioxide. When this combines with rainwater it produces sulfuric acid – an ingredient of "acid rain." Acid rain damages trees and lake life.

Acid rain slowly dissolves rocks such as limestone and marble.

LIMESTONE STATUES SUFFER EROSION BY ACID RAIN

EROSION IN ARID CLIMATES

IN HOT, DRY, DESERT areas extremes of temperature cause rocks to fragment. By day rock expands in the heat and by night it contracts in the cold. It is mainly physical weathering that occurs in arid climates, chiefly caused by wind. The sand-filled wind helps to erode rocks and build shifting sand dunes.

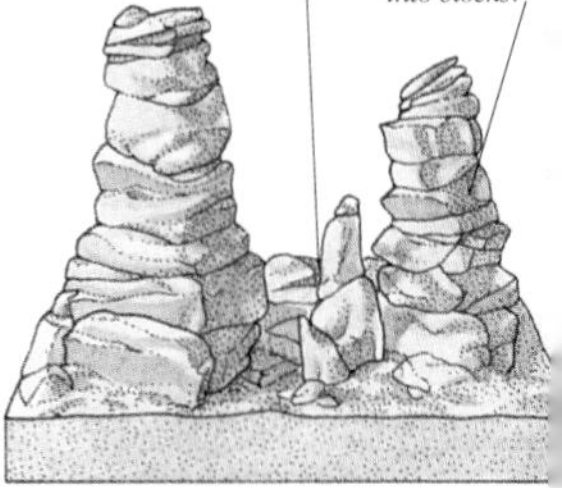
Some rocks break away and fall to the ground.
Larger rock masses split into blocks.

BLOCK DISINTEGRATION
Acute temperature changes can ca rocks to break up. Joints in the roc grow wider with the rock's cycle o expansion and contraction. Large pieces split into small blocks.

ONION-SKIN LAYERING
In the heat of the desert, a rock's surface may expand though the interior stays cool. At night, the surface of the rock cools and contracts. This daily process causes flaking on the surface of the rock, and the outer layers begin to peel and fall away.

SAND DUNES

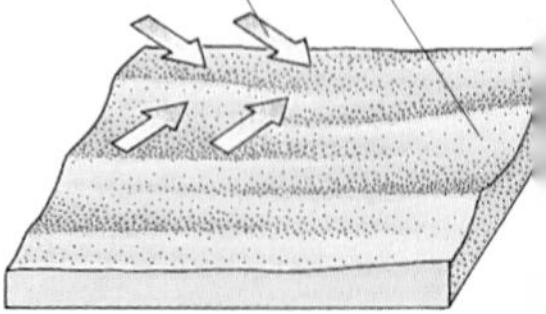
Direction of wind
Eddies build u the wind slo

LINEAR OR SEIF DUNE
This type of dune has long paralle ridges. It forms where the wind bl continually in one direction.

JGENS
ıd carried by the wind
lpts these strange forms
ed zeugens. Sand wears
ay soft rock leaving
ind areas of harder rock,
n into jagged shapes.

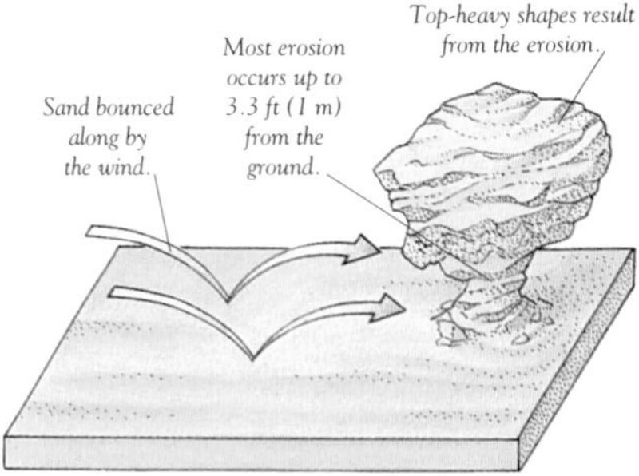

Sand bounced along by the wind.
Most erosion occurs up to 3.3 ft (1 m) from the ground.
Top-heavy shapes result from the erosion.

MUSHROOM ROCKS (PEDESTAL ROCKS)
The desert wind contains a great deal of sand which scours away the surface of rocks. Mushroom-shaped rocks are a result of this action. Rocks are worn away most at their base by the sand, leaving behind a landscape of rock pedestals.

TION THROUGH A
BARCHAN DUNE

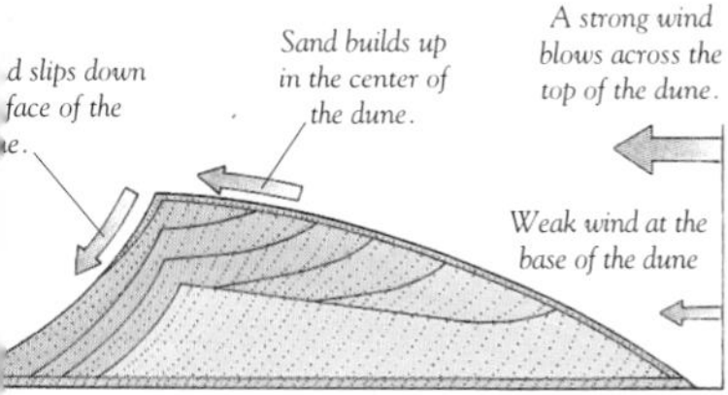

RCHAN DUNE
and dune with a cresent-shaped front and a
g, sloping rear is called a barchan dune. This
he most common dune shape for sandy deserts.

SAND DUNE FACTS

- Sand is composed mostly of the hard mineral quartz.
- Linear or seif dunes can reach 700 ft (215 m) high.
- Not all dunes are made of sand – dunes can form from salt crystals, gypsum, or shell fragments.
- Black sand dunes form in volcanic areas.

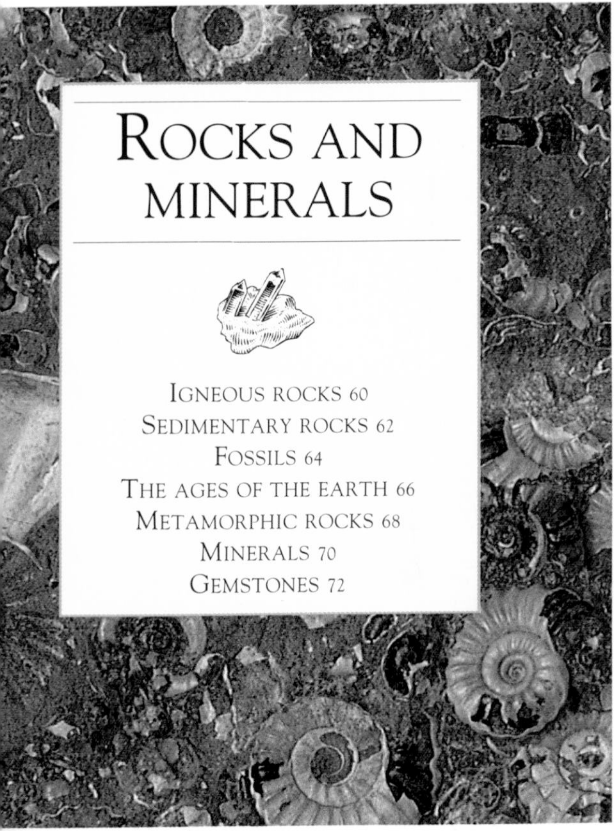

ROCKS AND MINERALS

IGNEOUS ROCKS

MAGMA THAT COOLS solidifies into igneous rocks. The rock material of the lower lithosphere and mantle is semimolten. Sudden release of confining pressure allows this material to change to liquid magma. Magma that cools and solidifies under the Earth's surface forms intrusive igneous rock. If it erupts as lava from a volcano and cools on the Earth's surface it forms extrusive igneous rock.

INTRUSIVE IGNEOUS ROCK
Sugar Loaf Mountain, Brazil forn
from magma that solidified unde
ground. Eventually, the surround
rock eroded, leaving this dome

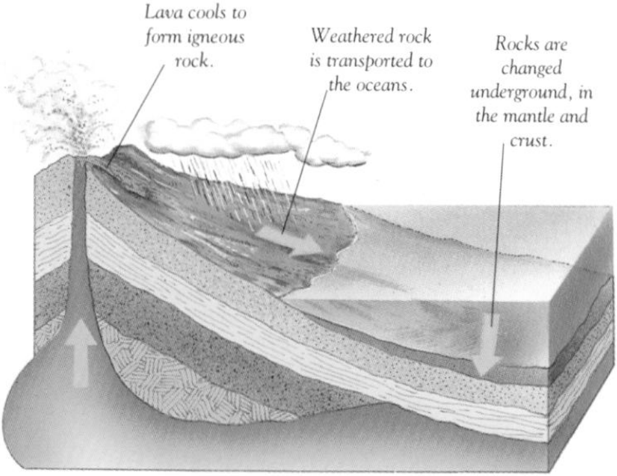

Lava cools to form igneous rock.
Weathered rock is transported to the oceans.
Rocks are changed underground, in the mantle and crust.

THE ROCK CYCL
Rocks constant
pass through a
recycling proce
Crustal movem
bring igneous r
to the surface.
rocks weather a
and the particl
build sediment
rocks. Pressure
heat undergrou
may change, or
metamorphose,
rocks before
they reemerge.

SYDNEY HARBOR BRIDGE
The supporting pylons of this famous bridge in Sydney, Australia are built from granite. The span of the bridge is 1,650 ft (503 m) and the arch is made of steel. Granite is often used as a building material because of its strength and availability.

IGNEOUS ROCK FACTS

- Basalt makes up most of the ocean floor.
- Obsidian was used in early jewelry and tools.
- Most continental igneous rocks are quartz, feldspar, and mica.
- Earth's first rocks were igneous rocks.

TYPES OF IGNEOUS ROCK

Shell-like, curved fracture

OBSIDIAN
This natural glass forms from rapid cooling of granitic lava.

Granite has coarse grains

GABBRO
This intrusive, coarse-grained rock forms from slow-cooled lava.

BASALT
Nonviscous lava that flows great distances forms fine-grained basalt.

PINK GRANITE
Granite is a common intrusive rock. Crystals of pink feldspar, black mica, and gray quartz minerals are visible.

SEDIMENTARY ROCKS

ROCK IS GRADUALLY weakened by the weather. Particles of rock are then carried off by rain or wind. These particles build up into layers of sediment. Evaporation of ground water may leave behind minerals, such as silica, calcium carbonate, and iron oxides, that cement the sediment grains (a process known as lithification). Studying sedimentary rock layers can reveal ancient environments.

STRATA
Over millions of years, laye of sediment are cemented i bands of rock called strata. Strata in the Grand Canyo Arizona preserve a record c the region's history.

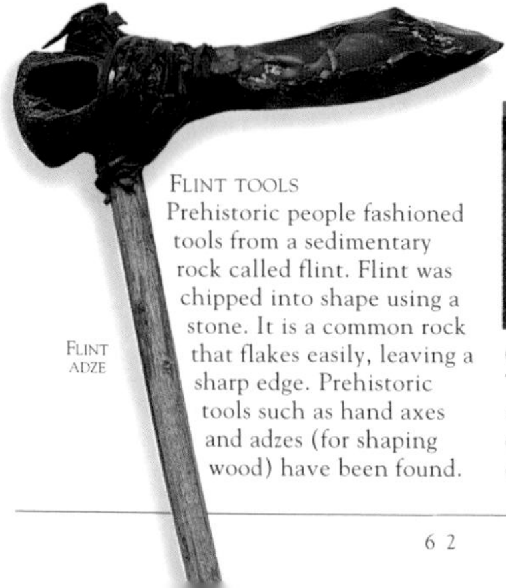

FLINT ADZE

FLINT TOOLS
Prehistoric people fashioned tools from a sedimentary rock called flint. Flint was chipped into shape using a stone. It is a common rock that flakes easily, leaving a sharp edge. Prehistoric tools such as hand axes and adzes (for shaping wood) have been found.

CHALK CLIFFS
These cliffs in Sussex, Englan are a type of limestone. They calcium carbonate (chalk) an contain fossils of microorganis

'PTIAN PYRAMIDS AT GIZA
ɔorate tombs (begun c. 2686) for the Egyptian raohs were constructed at Giza, Egypt. They e built from nummulitic limestone which tains many large marine fossils called *nummulites*.

SEDIMENTARY FACTS

- Chalk consists of tiny shells, visible only under a microscope.
- Mudstone forms from compressed mud grains, and sandstone from compressed sand grains.
- Oil is usually found in permeable and porous sandstones.

'YPES OF SEDIMENTARY ROCK

RECCIA
ngular fragments of ›ck are cemented ›gether to form breccia.

CHALK
Skeletons of tiny sea animals form this type of limestone. Chalk is fine-grained and soft.

RED SANDSTONE
Cemented sand grains coated with iron oxide make up this sedimentary rock.

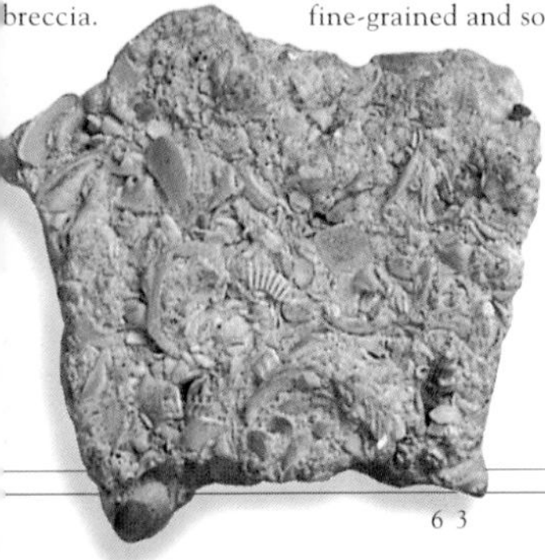

SHELLY LIMESTONE
This rock contains a great many fossils cemented together with calcite. Limestone usually forms in a shallow sea, though it can come from a freshwater environment. It is possible to find the source of a specimen by studying the fossils it contains.

FOSSILS

PLANTS AND ANIMALS that lived millions of years ago are preserved in rocks as fossils. A fossil is the remains of an organism, a cast of an animal or plant made from minerals, or even burrows or tracks left by animals and preserved in rock. Sedimentary rocks such as limestone or chalk hold fossils. Paleontologists are scientists who study fossils.

PLANT FOSSIL
Seed ferns like this one were widespread in the hot swamps of the late Carboniferous period. These primitive land plants, with some adaptations, still exist today.

THE FOSSILIZATION PROCESS

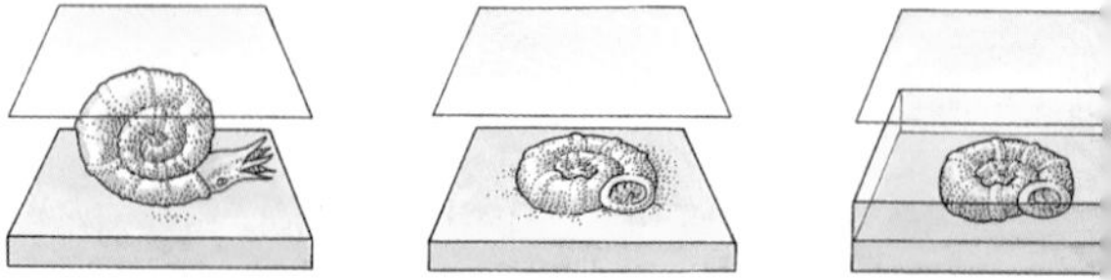

1 When an animal or plant dies underwater, it falls to the seabed. The soft parts of its body decay or are eaten by animals.

2 The organism is buried in layers of sediment. Hard parts of the animal, such as the shell, bones, or teeth, are preserved.

3 Minerals in the react with the shell to harden it. animals decay, leav space where a cast

SEED FERN
(*ALETHOPTERIS*)

INSECT FOSSIL
Early dragonflies, preserved in limestone, have been found in Europe and Australasia. This one dates from the Jurassic period.

DRAGONFLY
(*PETALURA*)

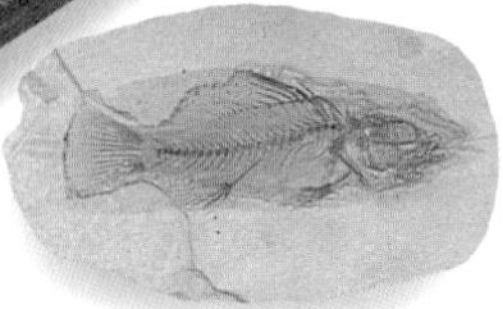

FISH FROM THE OLIGOCENE PERIOD

FISH FOSSIL
Fish are the most primitive vertebrates (animals with backbones). This fish first appeared 30 to 24.5 million years ago, long after the dinosaurs had died out.

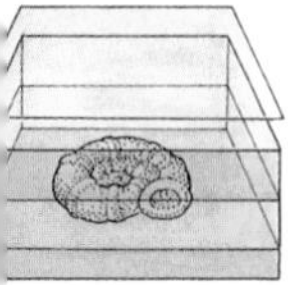

Further sediments cover the fossil. Uplift ıe lithosphere or ion may eventually ɔse the fossil.

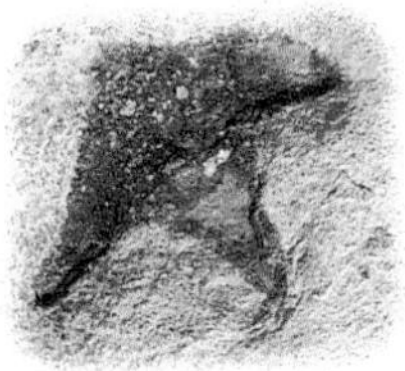

TRACE FOSSILS
Fossilized droppings or tracks are called trace fossils. This dinosaur footprint was left in mud 135 million years ago.

FOSSIL FACTS

- The earliest dinosaur, a *Herrerasaurus*, was found in Argentina in 1989 and dated at 230 million years old.
- The largest fossil footprint was left by a hadrosaurid. It is 4.46 in (1.36 m) long.
- Fossils of cells are the first evidence of life, 3.2 billion years ago.

THE AGES OF THE EARTH

THE LARGEST DIVISIONS of Earth's history are eras, periods, and epochs. The timescale is marked by the appearance of new life-forms. Life on Earth is never static – it constantly changes and evolves. Creatures become extinct and others appear. Some types of creatures may be short-lived and others survive unchanged for millions of years. Fossils can build up a picture of life in the past

JELLYFISH FOSSIL

PRECAMBRIAN FOSSIL
This fossil is about 570 million years old. It is a kind of primitive jellyfish that lived in Australia.

CARBONIFEROUS SWAMP
Extensive swamps covered Earth in the Mississipp and Pennsylvanian periods (363–290 million year ago). It was during this time that forests, containi seed plants and ferns, flourished. Some of these w preserved and now form coal deposits. The first reptiles and giant dragonflies lived in these swam

MORE FOSSIL FACTS

- Our species, *Homo sapiens*, first appeared about 40,000 years ago.
- A million Ice Age fossils were found preserved in the tar pits of La Brea, California.

DILOPHOSAURUS
WO-RIDGED LIZARD)

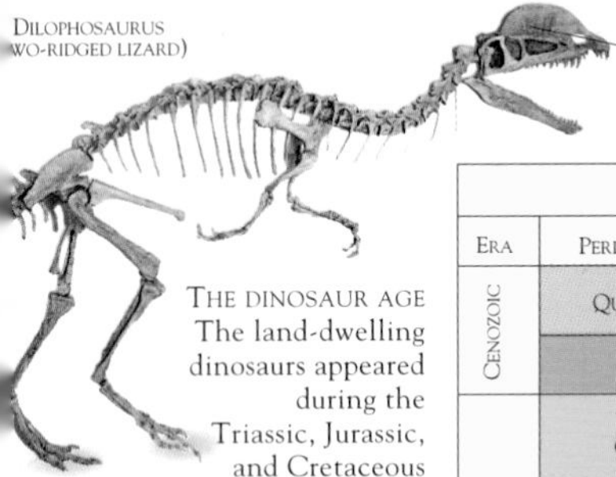

Distinguishing tall, double crest on the skull

THE DINOSAUR AGE
The land-dwelling dinosaurs appeared during the Triassic, Jurassic, and Cretaceous periods. This skeleton is from *Dilophosaurus*, an agile, predatory linosaur from the Jurassic period.

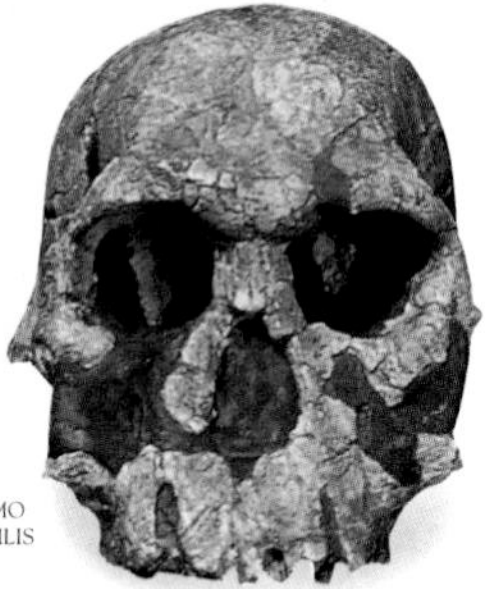

MO
BILIS

RLY HUMANS
mo habilis (handy man) is an ly human, dating from the aternary period. The name of s ancestor comes from the fossil dence that the early human was lled in using sophisticated tools.

GEOLOGICAL TIMESCALE	
ERA	PERIOD: MILLIONS OF YEARS AGO (MYA)
CENOZOIC	QUATERNARY (2 MYA–PRESENT)
	TERTIARY(65–2 MYA)
MESOZOIC	CRETACEOUS (146–65 MYA)
	JURASSIC (208–146 MYA)
	TRIASSIC (245–208 MYA)
PALEOZOIC	PERMIAN (290–245 MYA)
	PENNSYLVANIAN (320–290 MYA) MISSISSIPPIAN (363–320 MYA)
	DEVONIAN (409–363 MYA)
	SILURIAN (439–409 MYA)
	ORDOVICIAN (510–439 MYA)
	CAMBRIAN (570–510 MYA)
	PRECAMBRIAN (4,600–570 MYA)

SLATE MOUNTAINS
Fine-grained slate forms from sedimentary rocks. Rocks such as shale or mudstone are compressed during mountain building and changed into slate. Slate's aligned crystals let it split, or cleave, easily into flat sheets.

METAMORPHIC ROCKS

SEDIMENTARY, metamorphic, or igneous rocks are remade into new metamorphic rocks The rock doesn't melt but it is changed underground by pressure and heat. During mountain building, in particular, intense pressure over millions of years alters the texture and nature of rocks. Igneous rocks such as granite change into gneiss and sedimentary rocks like limestone into marble.

Carved white marble

MARBLE SCULPTURE
Michelangelo's statue of David is carved in marble. Marble comes in many varieties. It is a relatively soft rock that is often sculpted.

METAMORPHIC FACTS

- The oldest rock on Earth is a metamorphic rock. It is Amitsoc gneiss from Amitsoc Bay, Greenland.
- Rubies are found in metamorphic limeston in the Himalayas. The formed during mountain building.

GIONAL METAMORPHISM
treme pressure and heat created
ring mountain building lead to
;ional metamorphism.
etamorphism
this scale
n cover
ast
a.

Migmatite showing swirls of folded rock

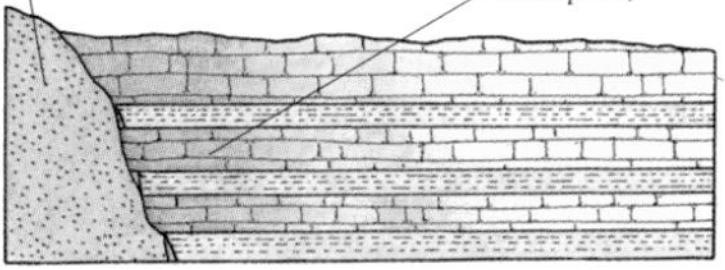

CONTACT METAMORPHISM
Rocks near to a lava flow or to an intrusion of igneous rock can be altered by contact metamorphism. This metamorphism affects a small area and is generated by heat alone.

TYPES OF METAMORPHIC ROCK

LATE
Aica crystals lie in the
ame direction in slate,
naking it easy to split.

SCHIST
Formed in moderate pressure and temperature conditions, schist often shows small, wavy folds.

ARBLE
Vhen limestone is
ubjected to intense
eat it becomes marble.

GNEISS
Igneous and sedimentary rocks can become gneiss. It forms at high temperatures.

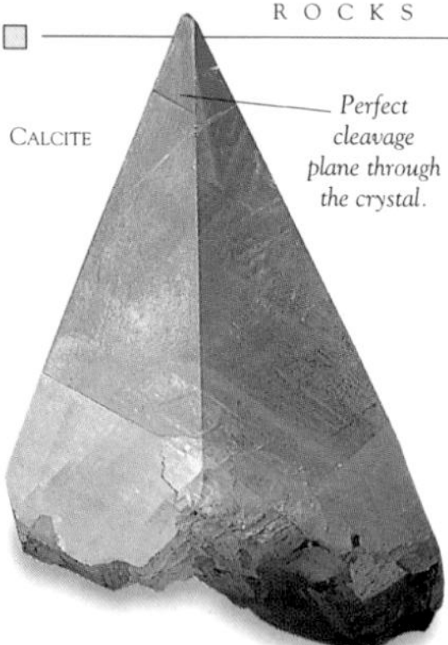

CLEAVAGE AND FRACTURE
Diamond and calcite cleave when they break. Cleavage is a smooth break between layers of atoms in a crystal. A fracture is an uneven break, not related to the internal atomic structure. Most minerals fracture and cleave.

MINERALS

ROCKS ARE MADE from non-living, natural substances called minerals, which may be alone or in combination. Marble is pure calcite, for example, but granite is a mixture of quartz, feldspars, and mica. Most minerals are formed from silicates (compounds of oxygen and silicon). Minerals with a regular arrangement of atom may form crystals. To identi a mineral, properties such as crystal structure, color, and hardness are tested.

MOHS' SCALE A German mineralogist named Friedrich Mohs devised a scale to compare the hardness of different minerals. A mineral is able to scratch any others below it on the scale and can be scratched by any mineral above it.	1	2	3	4

	TALC	GYPSUM	CALCITE	FLUOR

[M]INERAL FACTS

- Only a diamond will [sc]ratch a diamond.
- Quartz is found in [ig]neous, sedimentary, [an]d metamorphic rock.
- The word "crystal" [co]mes from a Greek [wo]rd *kyros* meaning [ic]y cold."

PLAGIOCLASE FELDSPAR

FELDSPAR
This type of rock-forming mineral is in both basalt and granite.

QUARTZ OR ROCK CRYSTAL

QUARTZ
A common mineral, quartz comes in many different colors. Amethyst and citrine are varieties of quartz.

[C]OLOR STREAKS

[S]cratching a mineral on an unglazed [t]ile produces a colored streak. The [c]olor of the powder left behind is [k]nown as the mineral's streak.

[O]RPIMENT - GOLDEN

HEMATITE - RED/BROWN

MICA
Found in metamorphic rocks such as schists and slate, flaky mica is also in igneous rocks like granite.

5	6	7	8	9	10
					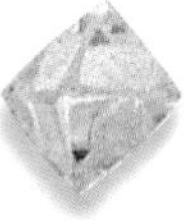
[APA]TITE	ORTHOCLASE FELDSPAR	QUARTZ	TOPAZ	CORUNDUM	DIAMOND

GEMSTONES

ONLY ABOUT 50 OF Earth's 3,000 minerals are valued as gemstones. Minerals such as diamonds, sapphires, emeralds, and rubies are commonly used as gems. They are chosen for their rarity, durability, color, and optical qualities. Gems may be found embedded in rocks or washed into the gravel of a river. Organic gemstones have a plant or animal origin. They include pearl, amber, and coral.

CROWN JEWELS
The British Imperial State Crown contains the Black Prince's ruby (in fact a 170-carat spinel) and the famous Cullinan II diamo

KIMBERLITE
Diamonds used to be found mainly in river gravels in India. In 1870 diamonds were discovered in volcanic rock, called kimberlite, in South Africa.

EMERALD
Green beryl crystals called emeralds contain chromium to make them green. Most emeralds are mined in Colombia. They have a hardness of 7.5 on Mohs' scale.

BRILLIANT
Skilled gem cutters, know as lapidaries, cut a rough crystal into a sparkling sto A diamond has 57 facets o faces ground onto it to ma it a brilliant.

NTHETIC GEMS
me gems can be roduced almost ıctly in a laboratory. ssolved minerals and oring agents stallize under strictly ıtrolled conditions produce perfect stals. Synthetic stals are used in dicine and the ctronics industry.

NATURAL RUBY

SYNTHETIC RUBY

GEMSTONE FACTS

- The Cullinan is the largest diamond ever found. It weighed 3,106 carats (1.37 lb/0.6 kg).
- About one in every 1,000 oysters contains a pearl, and about one in every 3,000 mussels.
- A carat equals 0.007 oz (0.2 g).

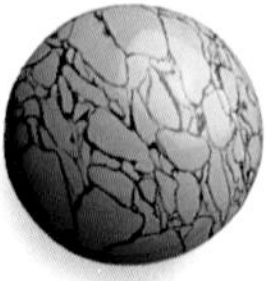

IMITATION TURQUOISE

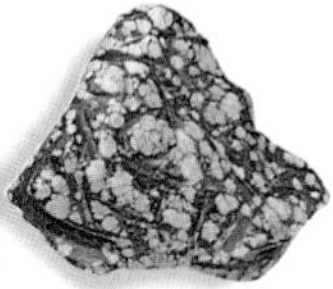

REAL TURQUOISE

IITATION GEMS
lass or plastic may be ed to imitate gems. he optical properties such imitations are fferent from those the genuine gem.

PEARL
Shellfish, such as mussels and oysters, grow pearls in their shells. When a grain of sand lodges in its shell, the animal covers it with nacre, a substance to stop irritation. This creates a pearl.

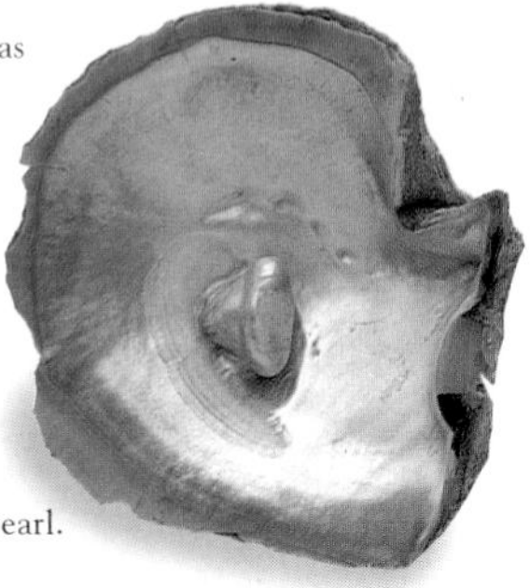

AMBER
Fossilized resin from coniferous trees is called amber. The trees that yielded this amber existed more than 300 million years ago and are extinct. Amber may contain insects trapped in the tree sap.

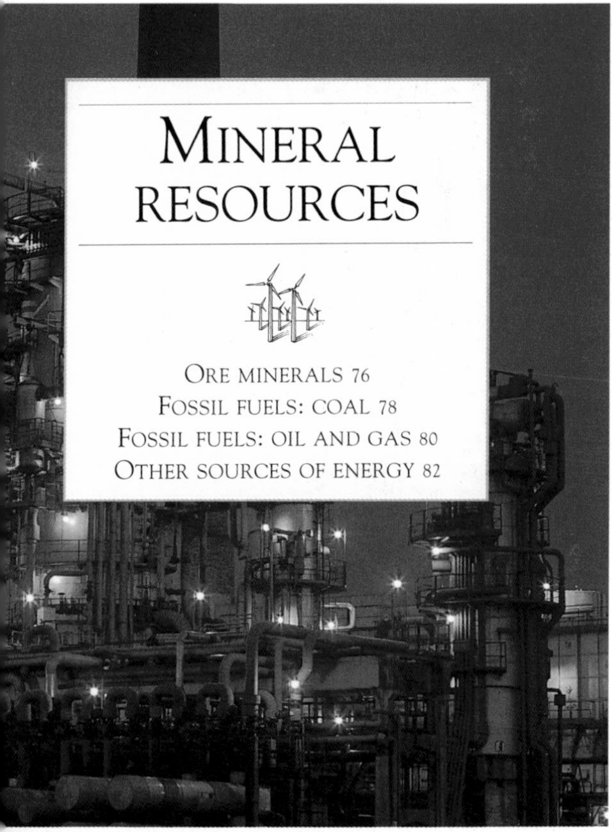

MINERAL RESOURCES

ORE MINERALS

A ROCK THAT yields metal in sufficient amounts is a metallic ore. Gold and copper can be found as pure metals, that is, uncombined with any other elements. Most other metals, such as iron and tin, are extracted from ores. Mined or quarried rock is crushed. The ore is separated and purified.

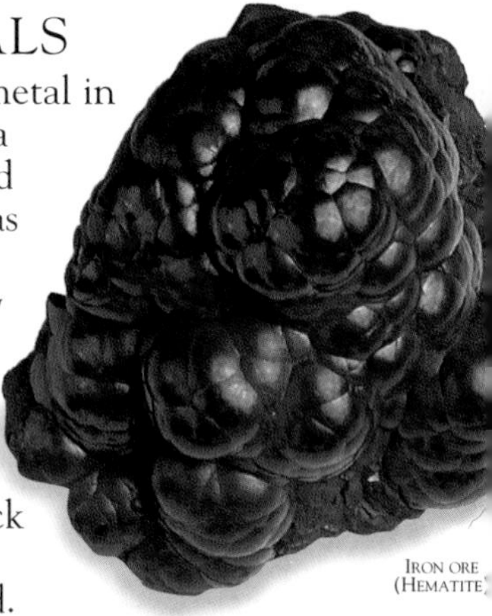

IRON ORE (HEMATITE)

ALUMINUM ORE (BAUXITE)

ALUMINUM FOIL

ALUMINUM

Lightweight aluminum is a good conductor of electricity and resists corrosion. It is extracted from its main ore (bauxite) by passing an electric current through molten bauxite and chemical fluxes.

GOLD FACTS

- The largest pure go nugget weighed 142.5 (70.9 kg). It was found in Victoria, Australia.
- 60 percent of the world's gold is mined in South Africa.
- Gold never loses its luster or shine.
- It is said that all th gold ever mined woul fit into an average fo bedroom house.

N

natite is an important
a ore. Iron can be cast,
ed, and alloyed with
er metals. Steel, used in
-building and industry,
roduced using iron.

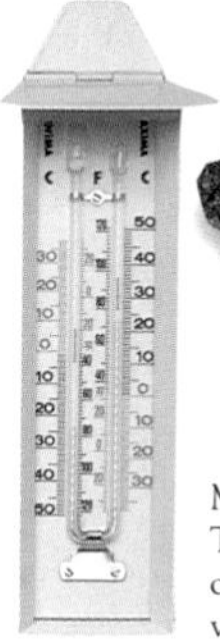

MERCURY THERMOMETER

MERCURY ORE (CINNABAR)

MERCURY

The primary mercury ore is called cinnabar. It is found near volcanic vents and hot springs, mostly in China, Spain, and Italy. Mercury is liquid at room temperature.

MINING

Blasting and boring rock in underground mines allows recovery of ores such as gold or tin. Dredging gravel or quarrying rock also retrieves ores.

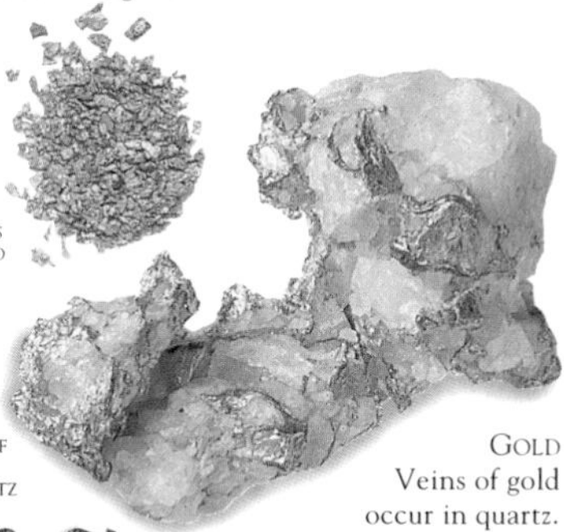

INS
OLD

S OF
LD
ARTZ

GOLD

Veins of gold occur in quartz. Panning or larger-scale dredging can retrieve gold grains from sand or river gravel deposits. About 1,500 tons (tonnes) of gold are produced each year.

GOLD GRIFFIN BRACELET

FOSSIL FUELS: COAL

PLANTS THAT GREW millions of years ago slowly changed to form coal. Vegetation in swamp areas, buried under layers of sediment, forms a substance called peat. Peat, in turn, is pressed into a soft coal called lignite. Soft, bituminous coal forms under further pressure. Anthracite is the hardest and most compressed coal. When coal burns, the energy of the ancient plants is released. Coal is used to fuel power stations that produce electricity. Coal supplies, like oil, are finite.

FROM PEAT TO COAL
Heat and pressure chan
crumbly brown peat int
hard black anthracite c

COAL FORMATION

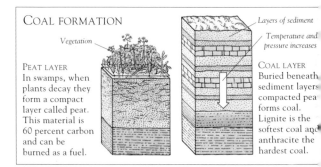

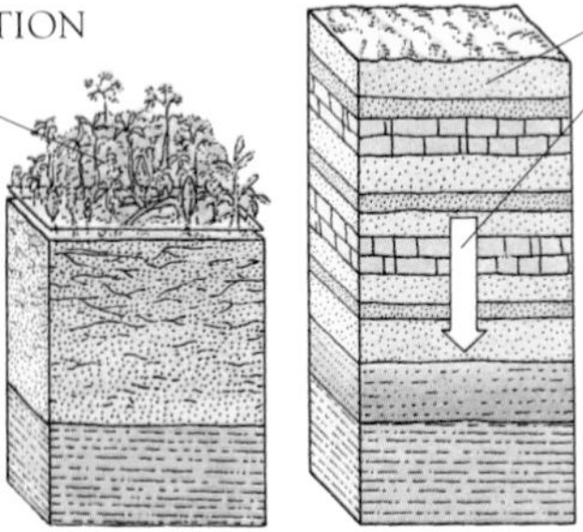

PEAT LAYER
In swamps, when plants decay they form a compact layer called peat. This material is 60 percent carbon and can be burned as a fuel.

COAL LAYER
Buried beneath sediment layers compacted pea forms coal. Lignite is the softest coal and anthracite the hardest coal.

INSIDE A COAL MINE

To reach a seam, or layer, of coal underground, rock must be blasted and bored away. Shafts go down from the surface to tunnels at different levels. Rock pillars and walls support the roof.

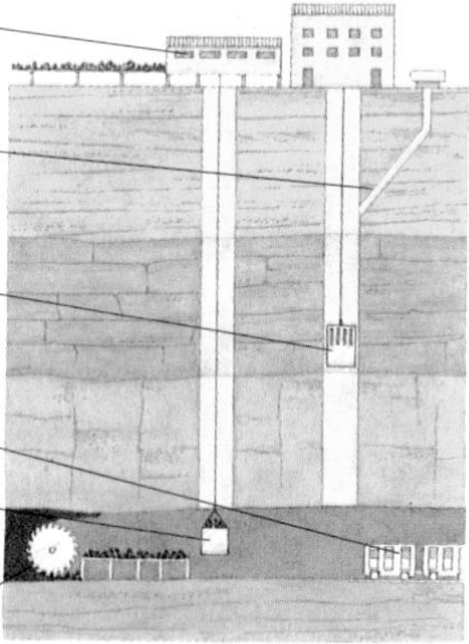

…L MINING

…ple have mined coal …e about 500 B.C. …ay's miners use drills … computer-controlled …hines. Special cutting …hines dig out the coal …e coal face. Deep coal …es deliver 2,000 tons …nes) of coal a day.

… OF COAL DEPOSITS

…mpy forests covered parts …urope, Asia, and North …erica, which were …lying during the …oniferous …od (360 – … million …s ago). …se tropical …t areas …ide most of …coal deposits … are mined today.

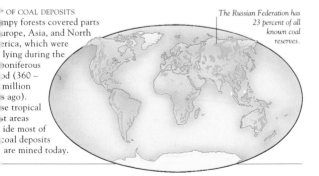

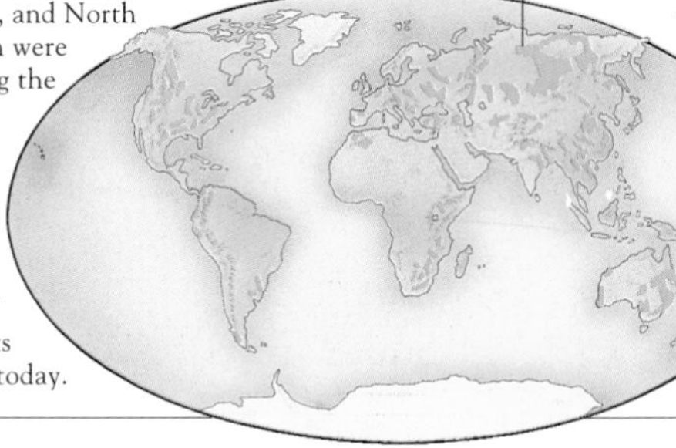

The Russian Federation has 23 percent of all known coal reserves.

FOSSIL FUELS: OIL AND GAS

MOST SCIENTISTS believe oil and gas are derived from the remains of ancient, single-cell marine animals. The organic remains were decomposed over millions of years to form oil and gas. The source rocks were compressed and heated to form liquid and gas hydrocarbons. Then they migrated to porous reservoir rocks that form oil and gas traps.

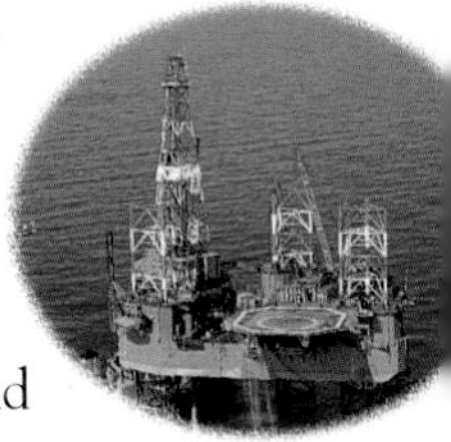

OIL RIG
An oil production platform floats but is tethered to the seabed. Oil is pumped up lc pipelines to the oil platforr

OIL AND GAS FORMATION

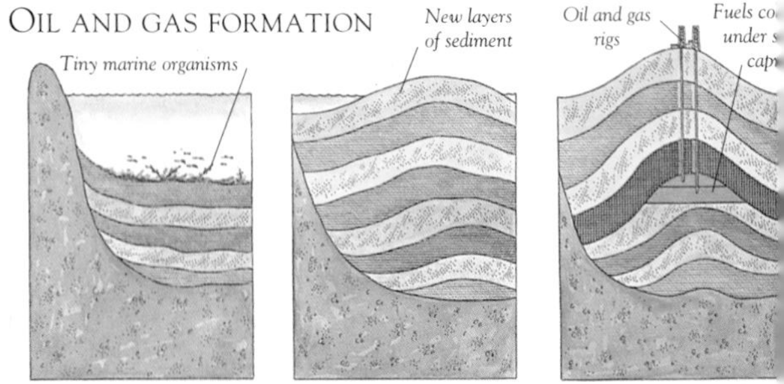

1 Decaying plants and animals sink to the sea floor. They lie buried by accumulating layers of sediment.

2 Heat and pressure increase as the sediments sink deeper. The organic remains become oil and gas.

3 Molecules of gas and oil rise through permeable rock and are held in porous rock.

RODUCTS MADE FROM OIL

.fter refining, oil can be ·parated to produce the ıbstances ethylene which forms plastics) nd ethanol. Ethanol is solvent used in paint ıanufacture. With xygen added, ethanol ıakes synthetic fibers.

SWEATER MADE FROM SYNTHETIC FIBERS

INFLATABLE TOY MADE FROM THIN PLASTIC

PAINT

PLASTIC TOY

OIL USED IN MACHINERY AND VEHICLES

P OF OIL AND GAS DEPOSITS

has been found in places such he Middle East and Arctic. North Central ıerica also e large and gas ds.

There are large oil deposits in the North Sea.

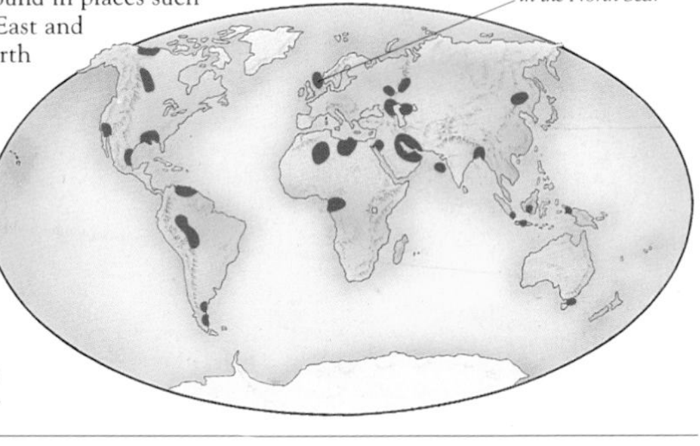

t gas fields ı the US, cially Alaska, in the Russian ration, usually to oil deposits.

OTHER SOURCES OF ENERGY

MOST OF THE energy the world uses for cooking, heating, or industry is produced by burning fossil fuels. These fuels cause pollution and will eventually run out. The Sun, wind, or water can be used to create pollution-free energy. This energy is renewable for as long as the Sun shines, the wind blows, and the tides rise and fall. In many places electrical energy is produced at nuclear power stations.

SOLAR ENERGY
The Sun's light energy captured by huge mirro
The energy is used to generate electricity.

TIDAL POWER
Inexpensive power can be generated in estuaries that have a large height difference between low and high tide, such as the Bay of Fundy in the US and Canada. Power can be generated both as the tide rises and as it falls.

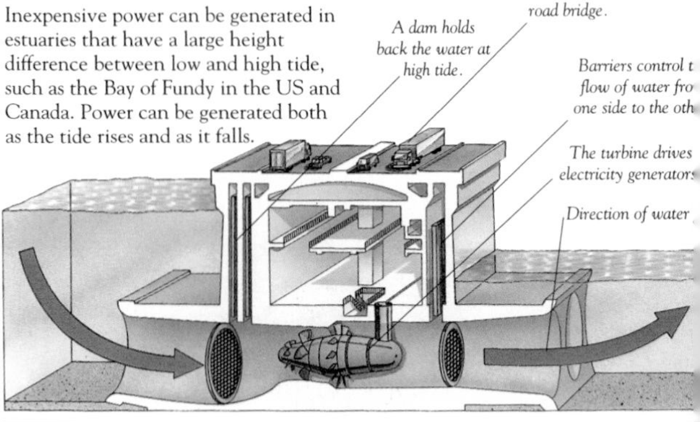
A dam holds back the water at high tide.
Dam acts as a road bridge.
Barriers control t flow of water fro one side to the oth
The turbine drives electricity generator
Direction of water

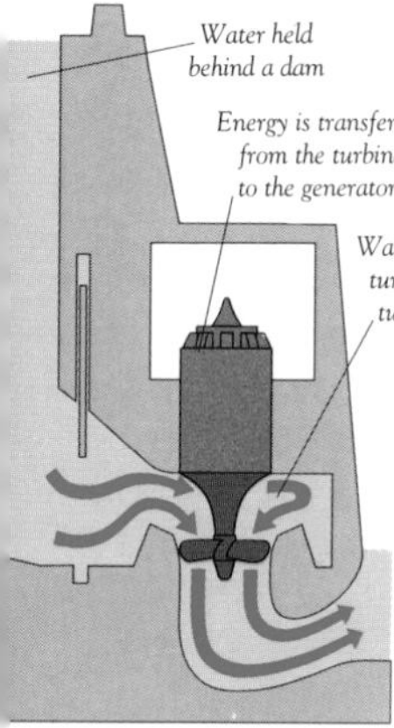
Water held behind a dam
Energy is transferred from the turbine to the generator.
Water flow turns the turbine

NUCLEAR ENERGY

Elements such as plutonium and uranium are used in nuclear power stations to generate energy. An atom of uranium can be split using a particle called a neutron. This produces heat and other neutrons. In turn, these neutrons split more atoms, generating further energy.

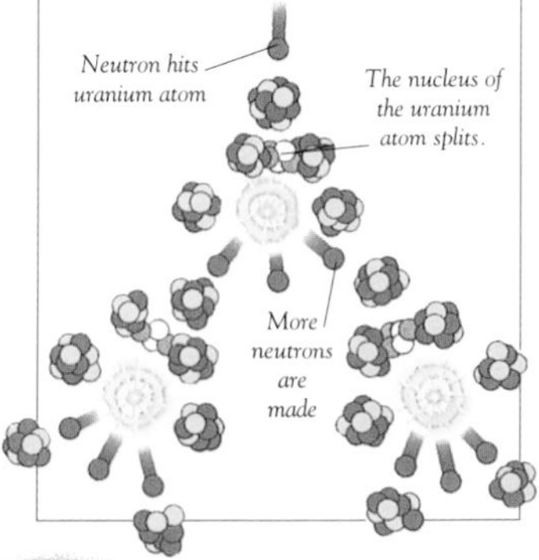
Neutron hits uranium atom
The nucleus of the uranium atom splits.
More neutrons are made

TER POWER

ydroelectric power station uses er power to produce electricity. e energy created by falling water s a turbine that turns an tric generator.

ERGY FACTS

- The first tidal power tion opened in La ance, France in 1966.
- Nuclear power first oduced electricity in e US in 1951.
- Hydroelectric power tions generate five rcent of all electricity.

WIND POWER

California has wind farms that contain thousands of windmills. Persistent winds spin the propellers, which drive electric generators. Each windmill can produce up to 300 kilowatts of power.

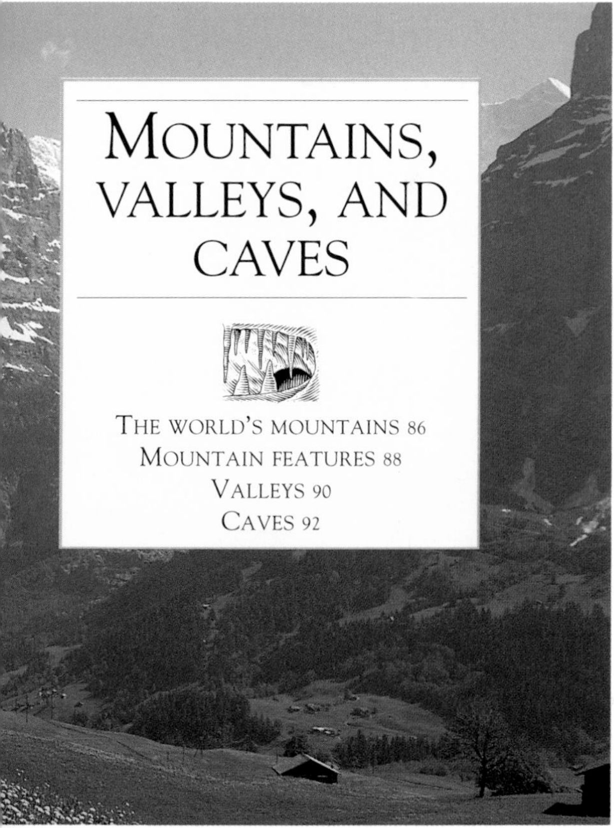

MOUNTAINS, VALLEYS, AND CAVES

THE WORLD'S MOUNTAINS

SOME MOUNTAIN ridges on land result from the collision of continental masses riding on tectonic plates. The Alps and Himalayas formed in this way. These lofty peaks continue to form as the Indian plate pushes the Eurasian plate. As the rocky plates fractur and crumple (or fault and fold) a mountain range takes shape. Ou planet has had several mountain-building episodes during its histor Some mountains are rising faster than they are being weathered aw

ANCIENT MOUNTAINS
The Scottish Highlands have been eroded into low rounded hills. These ancient mountains formed more than 250 million years ago.

YOUNG MOUNTAINS
Mountains such as the Himalayas continue to rise. The mountains are about 50 million years old and have jagged peaks.

TYPES OF MOUNTAIN

FAULT-BLOCK MOUNTAIN
When Earth's plates push into one another, faults or cracks in the crust appear. Huge blocks of rock are forced upward.

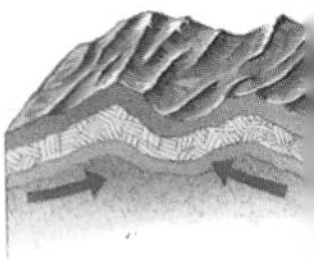

FOLD MOUNTAIN
At the meeting of t continents, the crus buckles and bends. The rocky crust is forced up into a mountain range.

)UNTAINS AND MOUNTAIN RANGES

ıe world's longest mountain ranges ıally follow the edges of the :th's plates.

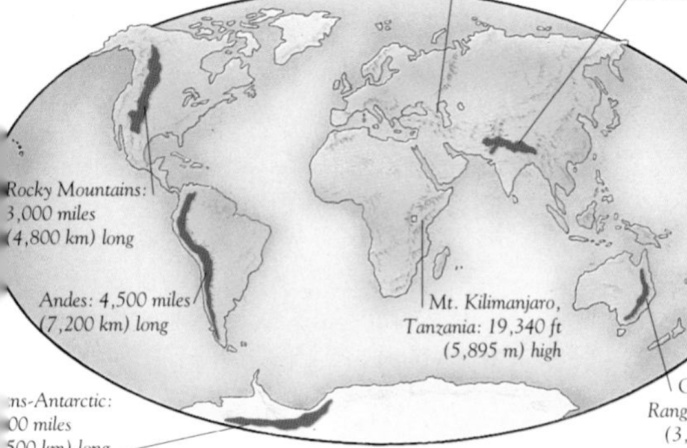

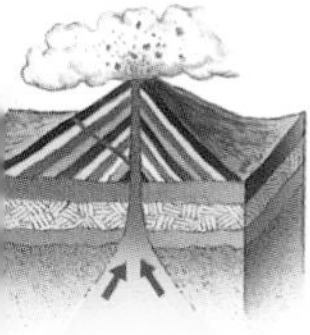

LCANO

va from a deep magma amber may erupt to m a volcano. A tall ne builds up from lava, , and rock ejected m the volcano.

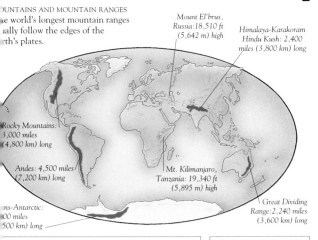

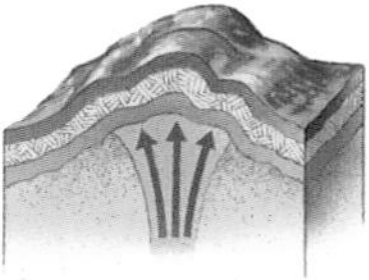

DOME MOUNTAIN

Rising magma forces up rocks near the surface. A dome-shaped mountain results.

MOUNTAIN FACTS

- The ten highest mountains on land are all in the Himalayas.
- Europe's Alps are part of a mountain belt that stretches from the Pyrenees in Europe to the Himalayas in Asia.
- The Himalayas grow at a rate of 3.3 ft (1 m) every 1,000 years.
- The Alps are the youngest of the world's great mountain ranges.

MOUNTAIN FEATURES

CONDITIONS ON MOUNTAINS can be harsh. As altitud increases temperature drops, air becomes thinner, an winds blow harder. Animal and plant life has adapte to survive in this environment. Mountains can be divided into several separate zones. The zones are similar whether the mountain lies in a tropical or temperate area and whether it is an isolated volcanic peak or part of a mountain range

MOUNT KILIMANJARO
Africa's tallest mountain is Mount Kilimanjaro in Tanzania. It is a solitary peak, not part of a range. In fact, it is a dormant volcanic cone. Despite lying near the equator, Kilimanjaro is permanently snow-capped.

WORLD'S HIGHEST MOUNTAINS PER CONTINENT

MOUNTAIN	CONTINENT	HEIGHT IN METERS	HEIGHT IN FEET
Mt. Everest, Nepal	Asia	8,848	29,028
Mt. Aconcagua, Argentina	South America	6,960	22,834
Mt. McKinley, Alaska	North America	6,194	20,320
Mt. Kilimanjaro, Tanzania	Africa	5,895	19,340
Mt. El'brus, Russia	Europe	5,642	18,510
Vinson Massif	Antarctica	5,140	16,863
Mt. Wilhelm, Papua New Guinea	Australasia	4,884	16,024

THE ANDES
The world's longest rang mountains on land is the Andes in South Americ The chain stretches for 4,500 miles (7,200 km).

JNDERSEA MOUNTAIN

Measured from the ocean floor, Mauna Kea, Hawaii is taller than Mount Everest. Rising 13,796 ft (4,205 m) above sea level, its base lies in a trough under the sea.

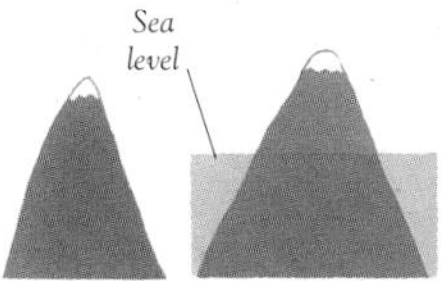

MORE MOUNTAIN FACTS

- On a mountain, the temperature drops 1.1°F (0.7°C) for every 330 ft (100 m) climbed.
- In warm Equatorial regions trees can grow at heights of 13,124 ft (4,000 m).

)UNTAIN VEGETATION

the altitude increases the ıperature falls. This effect ›duces distinct vegetation ł climatezones. The nt and animal life of :h zone varies. These the zones of the :opean Alps.

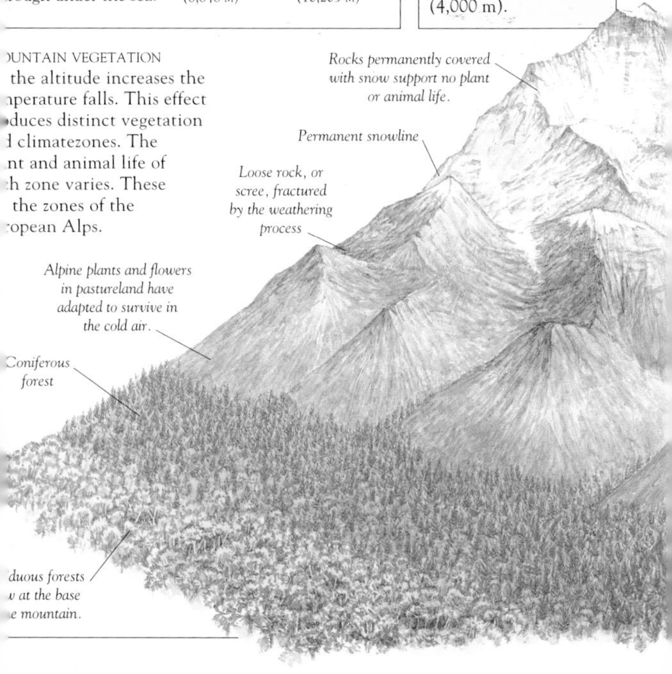

Rocks permanently covered with snow support no plant or animal life.
Permanent snowline
Loose rock, or scree, fractured by the weathering process
Alpine plants and flowers in pastureland have adapted to survive in the cold air.
Coniferous forest
duous forests v at the base e mountain.

VALLEYS

FORCES OF EROSION, especially water, control the shape of the landscape. Steep-sided valleys can be cut by fast-flowing mountain streams. Larger rivei wear a path through the land, shaping wide, flat valleys as th near the sea. Frozen water in glaciers also erodes rock, formi deep, icy gullies. Valleys sometimes form as a result of crustal movements that pull rocks apart at steep faults in the Earth's surface rocks.

GORGE
A ravine with steep sides is called a gorge. A canyon is similar to a gorge but it is usually found in desert areas.

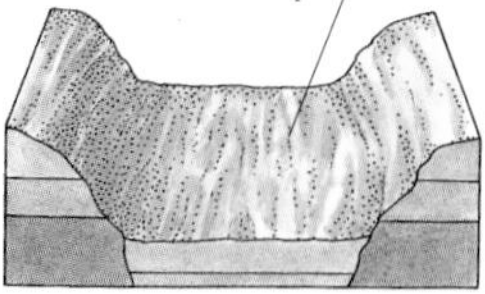
The lithosphere drops between the plate

RIFT VALLEY
Faults occur in the Earth's crust where two plates are moving apart. A long, straight valley, such as the African Rift Valley, forms between the faults.

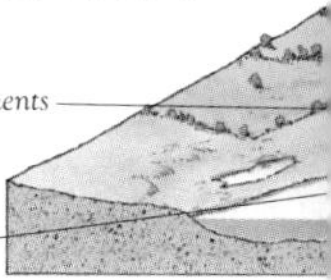
Rivers shed their sediments
A fan-shaped delta forms at the river mouth.

FJORD
Steep-sided estuaries such as those in Norway and New Zealand, are caused by glaciers deepening river valleys. As the ice melts and the sea level rises the fjords flood.

VALLEY FACTS

- Africa's Rift Valley stretches for 2,500 mile (4,000 km).
- The longest fjord, in Nordvest, Greenland, is 194 miles (313 km) lon

VER VALLEY FEATURES

the mountains, a
ɪt-flowing stream
ts through rock
ɛating a steep
ly. In the
ddle part of
e river, the
ɪter meanders
ross a broad
lley. Near the
, the river flows
ross a flat plain
it may fan
t into a
ta.

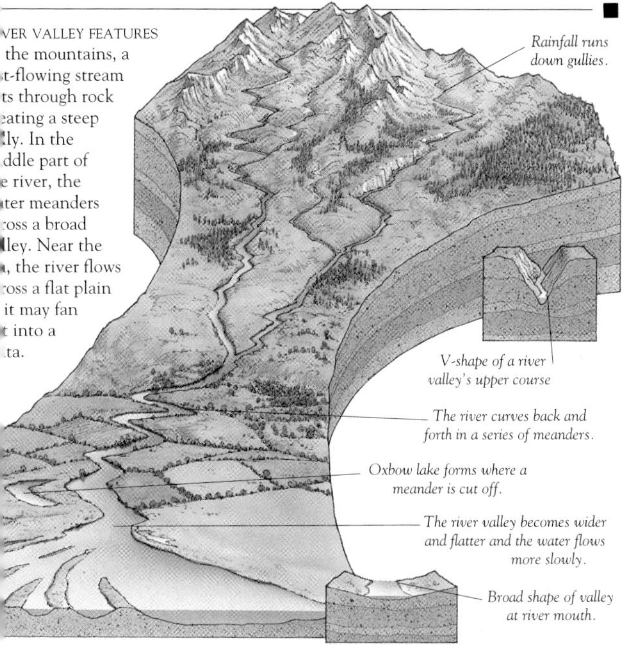

Rainfall runs down gullies.
V-shape of a river valley's upper course
The river curves back and forth in a series of meanders.
Oxbow lake forms where a meander is cut off.
The river valley becomes wider and flatter and the water flows more slowly.
Broad shape of valley at river mouth.

ADI

deserts it rains rarely, totaling
s than 10 in (250 mm) a year.
orrential downpour in such
arid area can cause a flash
od. The rain carves out a
annel called a wadi. Water
s down these dry river valleys
rying rocks and debris with it.

CAVES

UNDERGROUND CAVERNS and caves occur in several types of landscape. Different processes are responsibl for the development of caves. The action of ice, lava, waves, and rainwater cause subterranean openings. In particular, rainwater has a spectacular effect on limestone, producing vast caverns full of unusual shapes.

ICE CAVE
Beneath a glacier there is sometimes a stream of water that has thawed, called meltwater. The water can wear away an cave full of icicles in the glacier.

INSIDE A LIMESTONE CAVE
Carbonic acid in rain seeps into cracks in limestone and dissolves the rock. Underground tunnels and caves form as water dissolves the rock. Streams may flow down sinkholes in the rock into a cave system and emerge in another place.

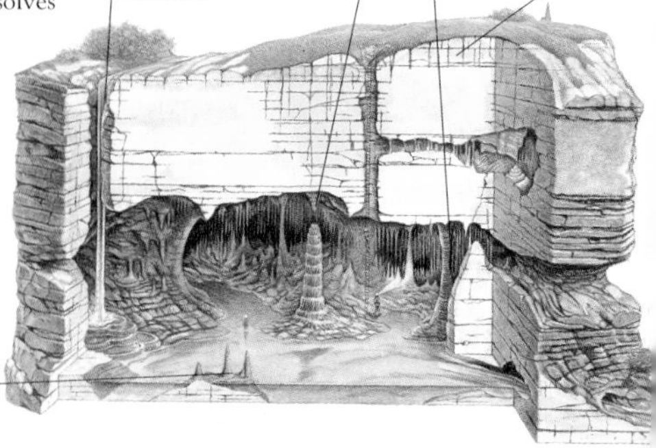
A stream may disappear down a sinkhole.
Stalactites hang down from the roof.
Columns form where stalactites and stalagmites j
Water seep cracks in r
Stalagmites grow upward from the cave floor.

LAVA CAVE
Some lava cools to form a thick crust. Below the crust a tube of molten lava flows. When it empties a cave remains.

CAVE FACTS

- Jean Bernard cave in France is the deepest in the world. It is 5,256 ft (1,602 m) deep.
- The longest stalactite – 20.4 ft (6.2 m) long – is in Co. Clare, Ireland.
- The tallest stalagmite is 105 ft (32 m) high. It is in the Czech Republic.
- America's Mammoth Cave system, Kentucky, is the world's longest cave system. It is 348 miles (560 km) long.

A CAVE
ıves may form at the base of ffs undercut by erosion. A ve can be worn through to m an arch. The top of this h may eventually llapse and leave an lated stack to be ffeted by the waves.

STALAGMITES AND STALACTITES

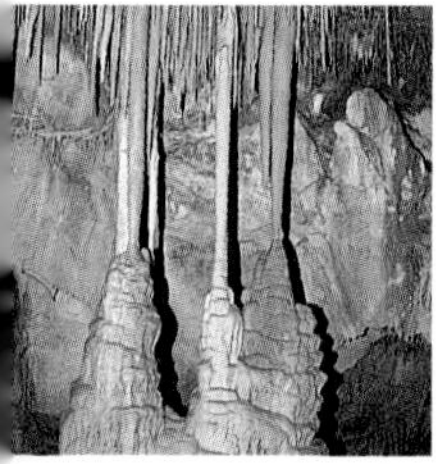

COLUMNS
Mineral deposits construct stalactites and stalagmites in caves. If the two shapes meet, they form a column.

STALACTITE
In limestone caves, calcite crystallizes from dripping water to create distinctive features, such as stalactites. These grow from the ceiling of the cave towards the floor.

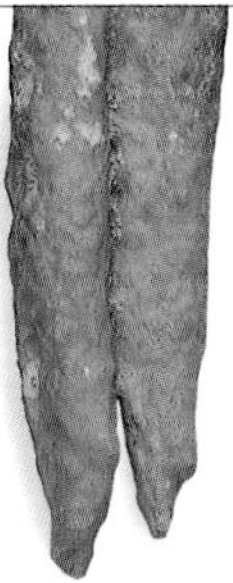

STALAGMITE
It may take several thousand years for a stalagmite to grow 1 in (2.5 cm) – drip by drip from a cave floor to the roof.

GLACIATION

THE WORLD'S GLACIERS

A GLACIER IS a mass of moving ice that originates in mountainous regions (mountain glaciers) or large, cold regions (ice caps). Ice builds up where there is more winter snow than melts in summer. The thick snow is compressed into ice. When the ice becomes very thick it begin to flow under its weight. Mountain glaciers are ice streams that follow former river valleys as they carry rock debris downhill.

GLACIAL DEBRIS
Rocks are smoothed when they are plucked up and carried along by a glacier. This rock has scratches, or striae, too.

THE WORLD'S LONGEST GLACIERS		
GLACIERS	LENGTH IN KM	LENGTH IN MILES
Lambert-Fisher Ice Passage, Antarctica	515	320
Novaya Zemlya, Russia	418	260
Arctic Institute Ice Passage, Antarctica	362	225
Nimrod-Lennox-King, Antarctica	289	180
Denman Glacier, Antarctica	241	150
Beardmore Glacier, Antarctica	225	140
Recovery Glacier, Antarctica	200	124

'ORE GLACIATION
e mountain valley
ved out by a river is
ally steep and shaped
e a letter V.

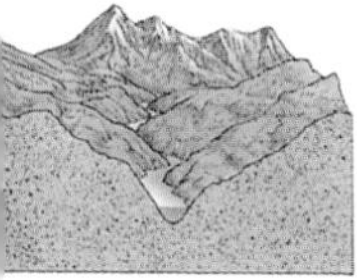

AFTER GLACIATION
A mountain glacier flows along the path of a river. The V-shape is eroded by the glacier into a U-shape.

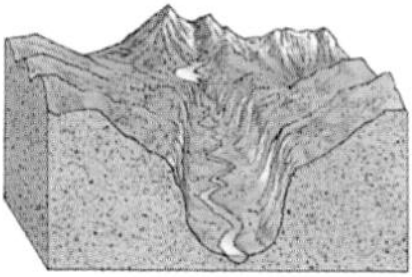

GLACIER FACTS

- Eight of the ten longest glaciers in the world are found in the Antarctic.
- About 10 percent of Earth's land surface is permanently glaciated.
- The fastest-moving glacier is the Quarayaq glacier in Greenland. It flows 65–80 ft (20–24 m) per day.
- Most glaciers move at a rate of about 6 ft (2 m) per day.

)SS-SECTION
A GLACIER

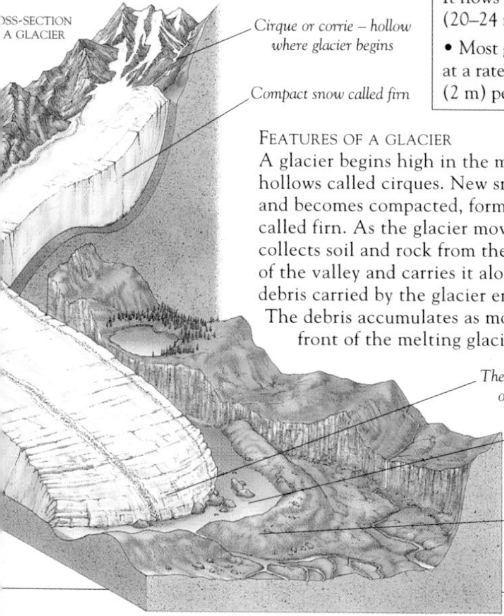

FEATURES OF A GLACIER
A glacier begins high in the mountains in hollows called cirques. New snow builds up and becomes compacted, forming denser ice called firn. As the glacier moves downhill it collects soil and rock from the floor and sides of the valley and carries it along. The rock debris carried by the glacier erodes the valley. The debris accumulates as moraine at the front of the melting glacier.

ICE CAPS AND ICE AGES

ANTARCTICA AND GREENLAND are blanketed in ice sheets up to 11,500 ft (3,500 m) thick. Many winter of snowfall accumulate to produce an ice cap, which eventually moves downhill as a broad glacial mass. In Earth's history, periods of extreme cold, called ice ages, brought glacial conditions as far south as Europe and North America. Our mild climate may only be an interval between ice ages.

ICE CAP
Vast ice sheets covering Antarctica and Greenland are known as ice caps.

FORMATION OF AN ICE CAP
Layers of snow build up during the winter months and become icy firn. Over several years, the result is a thick ice cap. Gravity pulls the ice down to the edges of the land.

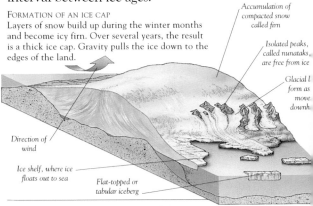

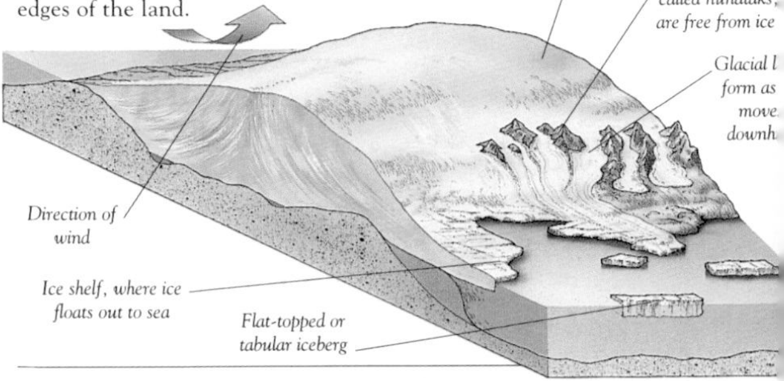

APH OF THE EARTH'S TEMPERATURE
e low points on the graph show the
e when the average earth
perature was cold enough to allow
or glacial advances.

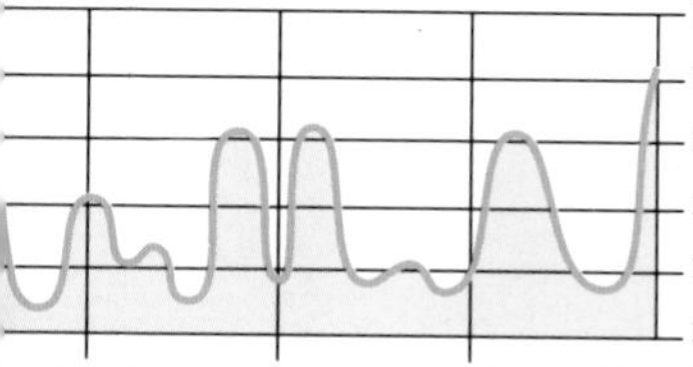

ICE FACTS

- Most of the world's freshwater – 75 percent of it – is stored in ice caps and glaciers.
- About 12 percent of the sea and 10 percent of the land is presently covered in ice.
- Almost 90 percent of the world's ice lies in the continent of Antarctica.

AGES
ing the last ice age,
ut 30,000 years ago, ice
ets covered a large part
he planet, particularly
th America and Europe.
ice ages are interspersed
warmer periods called
rglacials. Ice advanced
retreated with each
or temperature change.

PLEISTOCENE EPOCH – THE LAST ICE AGE

EXTENT OF ICE IN THE WORLD TODAY

SCIENTISTS EXAMINE A BABY MAMMOTH FOUND IN THE ICE

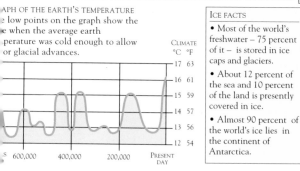

WOOLLY MAMMOTH

In Siberia, the remains of extinct animals called mammoths have been found in ice. They froze so quickly that their bodies were preserved virtually intact. These elephant-like mammals had curled tusks and woolly coats and lived during the last ice age.

AVALANCHES AND ICEBERGS

MOUNTAINS ARE inhospitable places. Winter snowstorms pile up layers of ice and snow. The layers may become unstable and swoop down the mountain in an avalanche, destroying anything in their path. Icebergs form when large chunks of ice break off (calve) from coastal glaciers or ice shelves. These are carried out to sea by ocean currents and are a hazard to ships.

SEA ICE
Seawater freezes when it reaches 28°F (–1.9°C). Sea ice is never more than abou 16 ft (5 m) thick. It can be used as a source of freshwat because the salt is left behi in the sea.

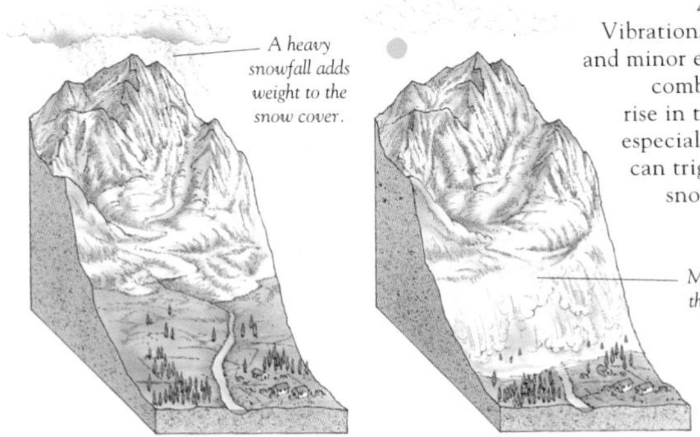
A heavy snowfall adds weight to the snow cover.
More snow j the avalanch it tumbles down the mountain

AVALAN
Vibrations from n and minor earthqua combined wi rise in temperat especially in spr can trigger a fa snow, calle avalan

ɔWLINE

ɜre is an elevation on a mountainside, led the snowline, below which ›w melts during summer. ove this elevation the ›w remains oughout the year. e snowline is her in more ıatorial ions.

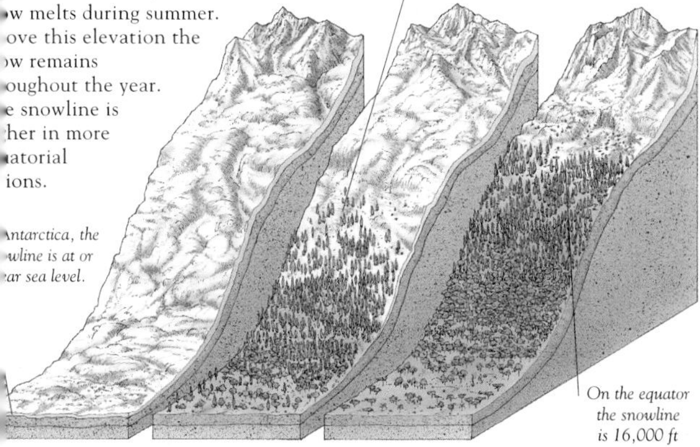

ʌntarctica, the
›wline is at or
ɛar sea level.
The snowline in the
European Alps is at about
9,000 ft (2,700 m) high.
On the equator
the snowline
is 16,000 ft
(4,900 m) high.

BERGS

ıciers and the floating edges of ice caps lose ınks of ice called icebergs into the sea, a cess known as calving. All icebergs are frozen hwater, rather than frozen seawater.

ICEBERG FACTS

- Only 12 percent of an iceberg can be seen above the ocean. 88 percent is under the water.
- The tallest iceberg was sighted in 1958 off Greenland. It was 550 ft (167 m) high.
- The largest iceberg, spotted in the Pacific Ocean in 1956, had an area of 12,500 miles^2 (32,500 km^2).

OCEANS, ISLANDS, AND COASTS

THE WORLD'S OCEANS

SEEN FROM SPACE, the Earth looks blue and watery. This is because two-thirds of it is covered with water. The water is held in oceans and seas. (Seas are surrounded by land.) There are five oceans: three are in the Southern Hemisphere. Major currents circulate the oceans counter-clockwise in the Southern Hemisphere and clockwise in the Northern Hemisphere.

THE WORLD'S LARGEST OCEANS AND SEAS

OCEAN OR SEA	AREA IN KM²	AREA IN MILES²
Pacific Ocean	166,229,000	64,181,000
Atlantic Ocean	86,551,000	33,417,000
Indian Ocean	73,422,000	28,348,000
Arctic Ocean	13,223,000	5,105,000
South China Sea	2,975,000	1,149,000
Caribbean Sea	2,516,000	917,000
Mediterranean Sea	2,509,000	969,000
Bering Sea	2,261,000	873,000

FORMATION OF OCEANS

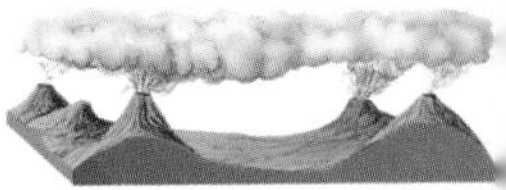

THE ATMOSPHERE FORMS
The semimolten surface of the Earth was covered by volcanoes. Hot gases and water vapor emitted by volcanoes formed the Earth's early atmosphere.

THE RAINS FALL
The water vapor in this early atmosphere condensed as rain. Rainstorms poured down on the planet and filled the vast hollows on the Earth's surface.

THE OCEANS FORM
These huge pools became the oceans. The water was hot and acidic. Later, plant life evolved and modified the chemical composition of the atmosphere and oceans.

:EAN ZONES

eanographers split
: oceans into zones
:ording to depth.
ly near-surface
ters are sunlit.
nerally the coldest
ter is at the bottom
the ocean. Water
ssure increases with
oth, but sea
atures have adapted
the different zones.

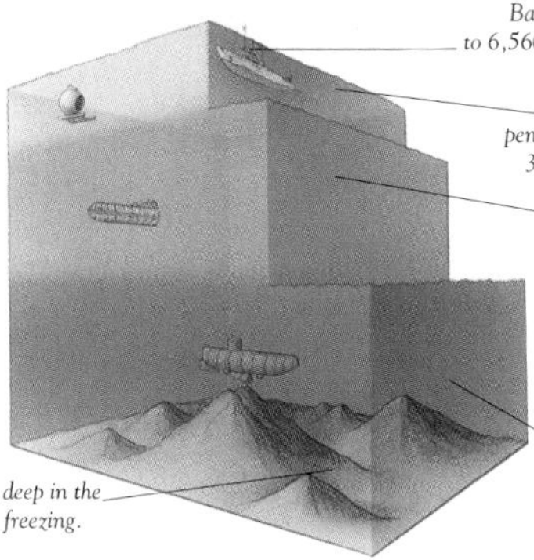

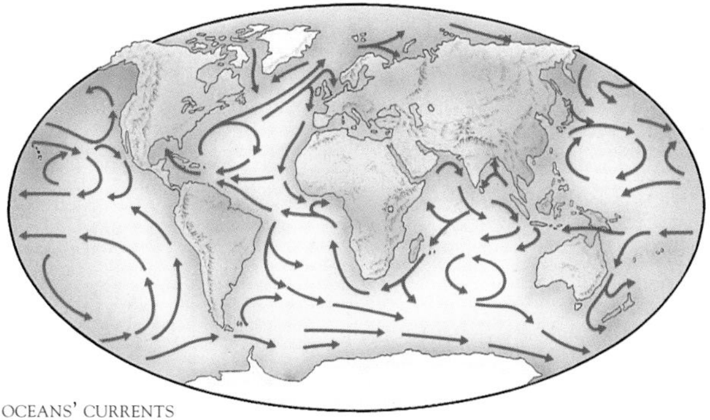

E OCEANS' CURRENTS

rrents may be warm or cold. They flow across the
face or deep beneath it. The wind controls surface
rents, which flow in circular directions. Currents
ry some of the Sun's heat around the planet,
rming polar areas and cooling tropical areas.

KEY	
COLD CURRENT	→
WARM CURRENT	→

WAVES AND TIDES

THE OCEANS AND seas are always moving. Buffeted by the wind and heated by the Sun, waves and currents form in the oceans. Ripples on the surface of the water may grow into waves that pound the shore and shape the coastlines. The Moon's and Sun's gravity pull the oceans, causing a daily and monthly cycle of tides.

WHIRLPOOL
An uneven channel can cause several tidal flows to collide. The currents surge upwards and rush into eacl other. Eddies and whirlpoc form on the surface.

Rotation of the Earth balances the Moon's gravitational pull.

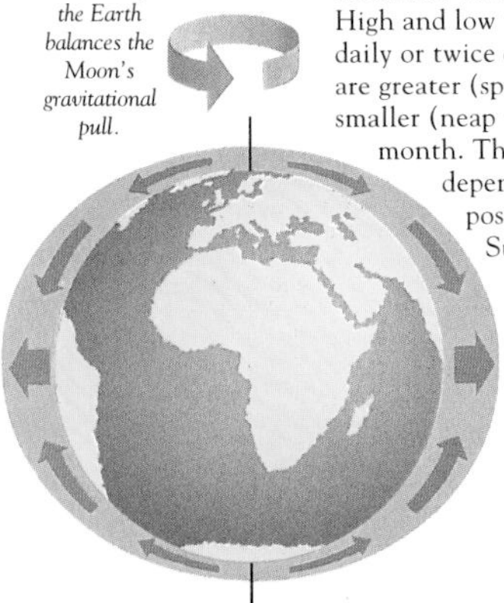

Earth's spinning on its axis affects the tides.

MONTHLY TIDES
High and low tides occur on a daily or twice daily rhythm. Tides are greater (spring tides) or smaller (neap tides) twice each month. This tide range depends on the relative positions of the Moon, Sun, and Earth.

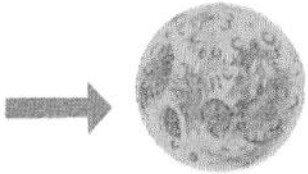

The Moon's gravity pulls the oceans.

The oceans on each side of the Earth rise approximately every 12 hours and then fall back.

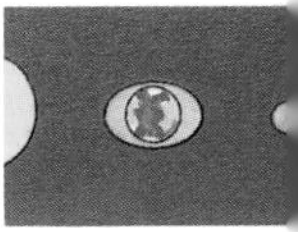

SPRING TIDES
The alignment of th Sun, Earth, and Mo create spring tides.

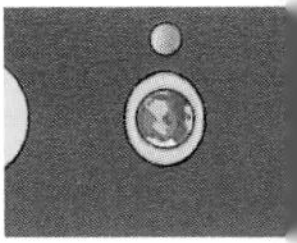

NEAP TIDES
Opposing pulls of Sun and Moon cau neap tides.

ILF STREAM

:urrent of warm water called the lf Stream moves from the Gulf Mexico across the Atlantic, nging mild winter weather to western coasts of Europe. :e a huge river at sea, the lf Stream flows 100 miles 0 km) a day. This current gyre is 37 miles (60 km) wide 1 2,000 ft (600 m) deep. As Gulf Stream nears Europe it is led the North Atlantic Drift.

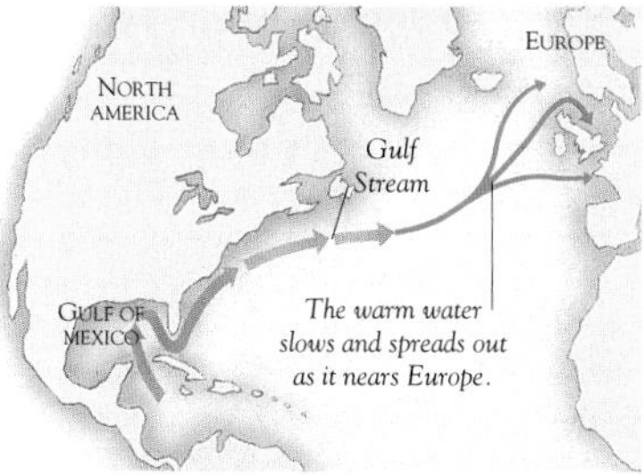

EUROPE
NORTH AMERICA
Gulf Stream
GULF OF MEXICO
The warm water slows and spreads out as it nears Europe.

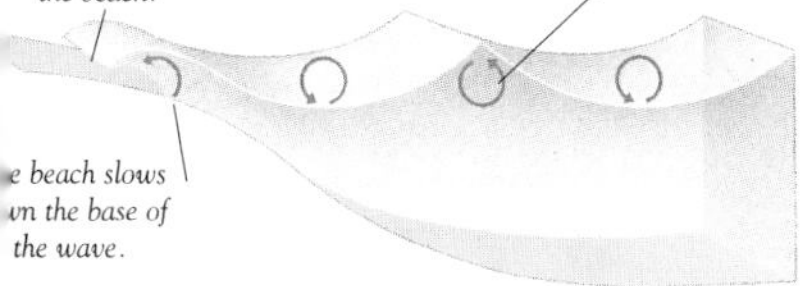

: top part of the wave, : crest, continues up the beach.
Particles near the surface turn over and over
e beach slows vn the base of the wave.

WAVE MOVEMENT

Waves travel toward the shore from large storms at sea. But it is not the water particles that travel, only the wave form. The particles rotate as each wave passes and return to their original position.

'AVE FACTS

The highest recorded ave was seen in the estern Pacific. It was 2 ft (34 m) high from ough to crest.

The Antarctic ircumpolar Current ws at a rate of 3 billion ft^3 (130 llion m^3) per second.

Hawaiian tides rise : in (45 cm) a day.

COMPOSITION OF SEAWATER

The oceans contain dissolved minerals, some washed from the land by rivers. The predominant constituents of seawater are sodium and chloride which together form salt. About 3.5 percent of the weight of ocean water is salt.

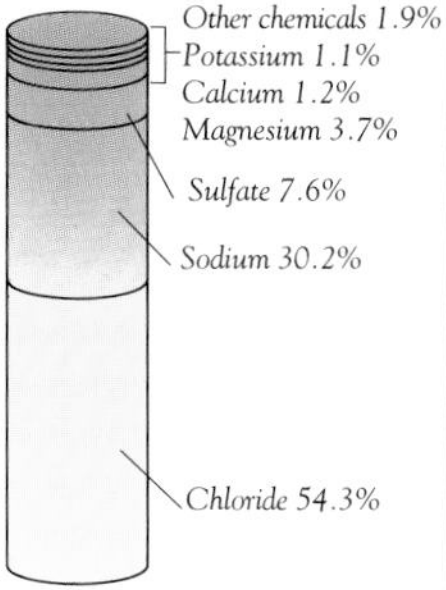

THE OCEAN FLOOR

THE WORLD UNDER the oceans has both strange and familiar features. Similar to a landscape on dry ground, mountains, valleys, and volcanoes dot the ocean floor. Once scientists had equipment to explore the ocean bed, they discovered that tectonic plate movement had caused many ocean floor features, including trenches, seamounts, and submarine canyons.

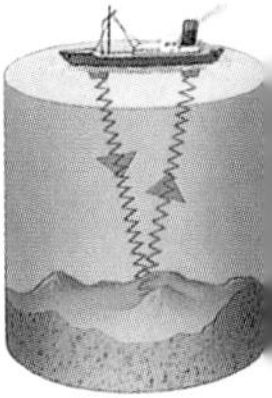

MAPPING THE OCEAN
Oceanographers use precision echo-sounding, which bounces signals off the ocean bed, to map the ocean floor's contours

FEATURES OF THE OCEAN FLOOR

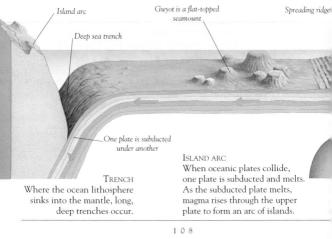

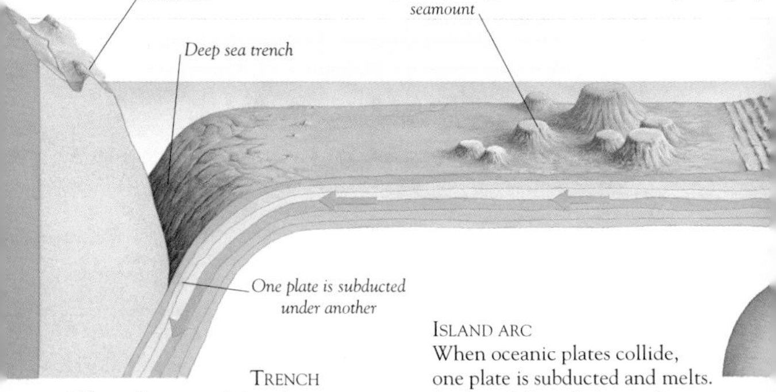

TRENCH
Where the ocean lithosphere sinks into the mantle, long, deep trenches occur.

ISLAND ARC
When oceanic plates collide, one plate is subducted and melts. As the subducted plate melts, magma rises through the upper plate to form an arc of islands.

EP-SEA EXPLORATION		
DIVERS AND VEHICLES	DEPTH IN METERS	DEPTH IN FEET
nge diver holding breath	15	49
'SCUBA sports diver	50	164
rig divers with diving bell	250	820
epest experimental dive	500	1,640
ton's benthoscope	1,370	4,494
isteau diving saucer	3,350	11,000
nkai submersible	6,500	21,325
este bathyscape	10,911	35,800

BATHYSCAPE
The deepest dive was made by the bathyscape *Trieste* in 1960. It dived in the world's deepest trench, the Mariana Trench.

AMOUNT
underwater volcano t rises over 3,280 ft 000 m) is a seamount.

ABYSSAL PLAIN
This sediment-covered plain lies at a depth of about 15,000 ft (4,000 m).

SUBMARINE CANYON
Dense sediment flows from the continental shelf have eroded deep canyons.

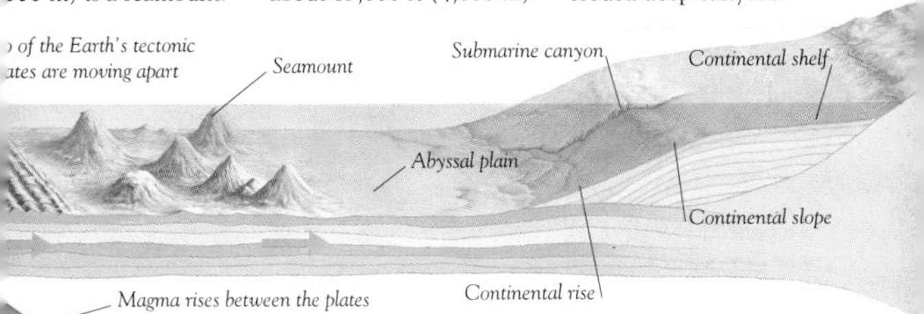

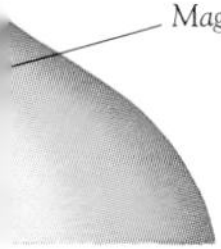

OCEANIC RIDGE
gma rising between two onic plates forms a ridge.

CONTINENTAL SLOPE
The continental slope is the edge of the continent. It descends from the edge of the continental shelf to the rise, or into a trench.

CONTINENTAL SHELF
Stretching from the edge of the land like a vast plateau under the sea, the continental shelf averages 43 miles (70 km) wide. The ocean here is about 1,600 ft (250 m) deep.

OCEAN FEATURES

OCEANS ARE a riçh source of many useful mineral substances. Seawater contains nutrients for phytoplankton and both metallic and nonmetallic minerals. Deposits of oil and gas are found in continental shelf sediment layers where tectonic paltes are rifting apart.

BLACK SMOKERS

In 1977, scientists discovered strange chimneys, formed from minerals on the ocean floor, called black smokers. In rift valleys between spreading ridges, they eject water as hot as 572°F (300°C), containing manganese and sulfur.

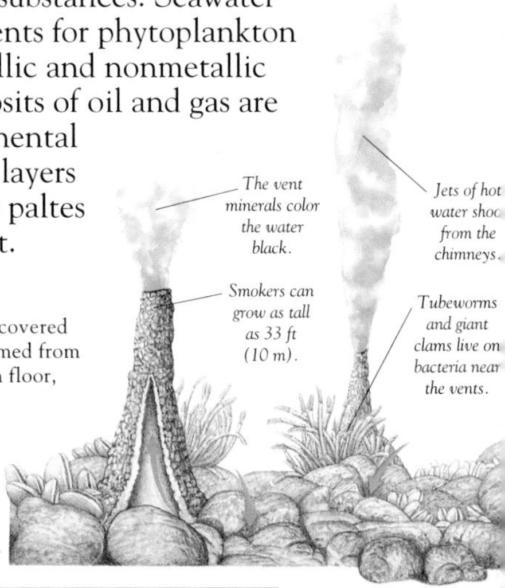
The vent minerals color the water black.
Jets of hot water shoo from the chimneys.
Smokers can grow as tall as 33 ft (10 m).
Tubeworms and giant clams live on bacteria near the vents.

THE WORLD'S DEEPEST SEA TRENCHES		
TRENCH	DEPTH IN METERS	DEPTH IN FEET
Mariana Trench, West Pacific	10,920	35,827
Tonga Trench, South Pacific	10,800	35,433
Philippine Trench, West Pacific	10,057	32,995
Kermadec Trench, South Pacific	10,047	37,963
Izu-Ogasawara Trench, West Pacific	9,780	32,087

OCEAN PRODUCT FACTS

- There are 0.000004 parts per million of gol in the ocean.
- It takes a million years for a manganese nodule to grow 0.08 in (2 mm) in diameter.

)CEAN PRODUCTS

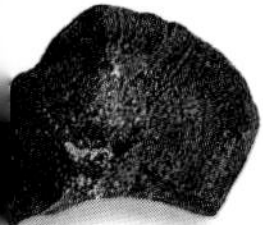

ANGANESE
odules from the
abed are used
industry.

OIL
This is a non-renewable fossil fuel. It is pumped from rocks in the continental shelf.

ND
ɔck pounded by
ıves becomes
nd. In volcanic
eas it is black.

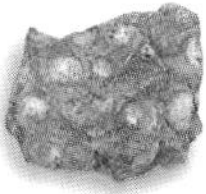

CORAL
Like sand, coral is found in coastal waters.

Diamonds in gravel are known as alluvial diamonds.

DIAMONDS IN GRAVEL
Off the coasts of Africa and Indonesia, diamonds can be found in continental shelf gravels. Most have been washed down by rivers into the sea.

EAN FLOOR SEDIMENT
continental shelf is covered with
l, mud, and silt washed onto it from
rs. In the deep ocean, the floor is
ed with ooze. This contains the
ains of dead marine life.

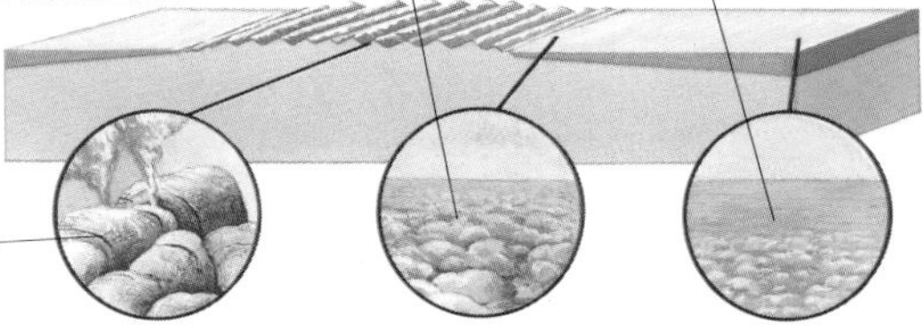

Rock is carried 311 miles (500 km) from the ridge over 5 million years. Sediment gathers.
After 10 million years the rock has moved farther from the ridge. It is now covered with thick sediment.
rock erupted
the mantle
nid-ocean
es has no
nent cover.

ISLANDS

A PIECE OF LAND smaller than a continent and surrounded by water is called an island. Magma rising from volcanic vents in the oceanic lithosphere creates islands in the sea. An arc of islands appears where a tectonic plate is subducted. Some islands exist only when the tide is high; at low tide it is possible to walk to these islands. Small islands may exist in rivers and lakes. In warmer regions coral reefs may grow from the sea, built by living organisms.

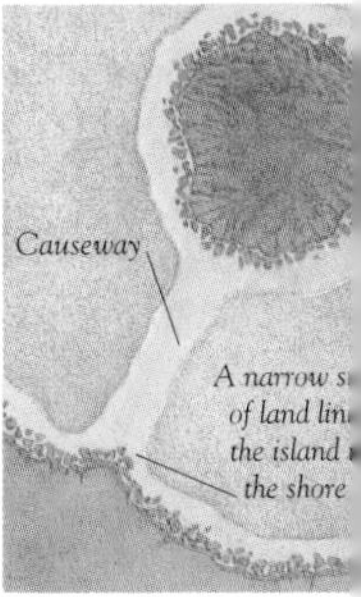

CAUSEWAY
A change in sea level can create an island. Land may be accessible only at low tide by a causeway. At high tide the island is cut off.

ISLAND FACTS

- Bouvet Island is the most remote island – about 1,056 miles (1,700 km) from the nearest landmass (Antarctica).
- Kwejalein in the Marshall Islands, in the Pacific Ocean is the largest coral atoll. Its reef measures 176 miles (283 km) long.

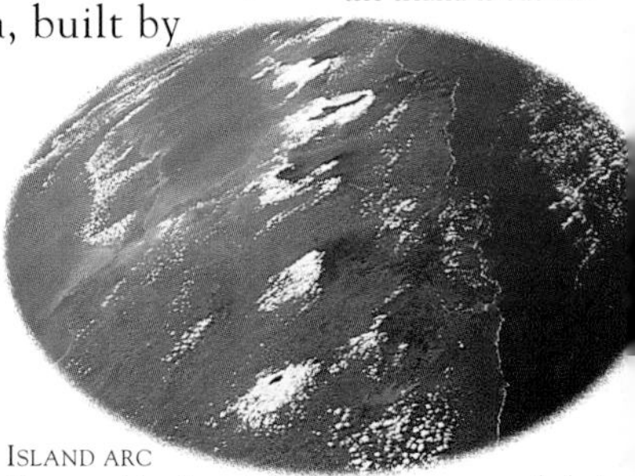

ISLAND ARC
On one side of a subduction zone, a curved chain arc of volcanic islands may be pushed up from un the ocean floor. From space the numerous volcan peaks on the islands of Indonesia are clearly visib

ɪE WORLD'S LARGEST ISLANDS

Island	Area in km ^{2}	Area in miles ^{2}
enland	2,175,219	839,852
v Guinea	792,493	305,981
neo	725,416	280,083
lagascar	587,009	226,644
'in Island, Canada	507,423	195,916
ıatra , Indonesia	427,325	104,990
ıshu, Japan	227,401	87,799
at Britain	218,065	84,195

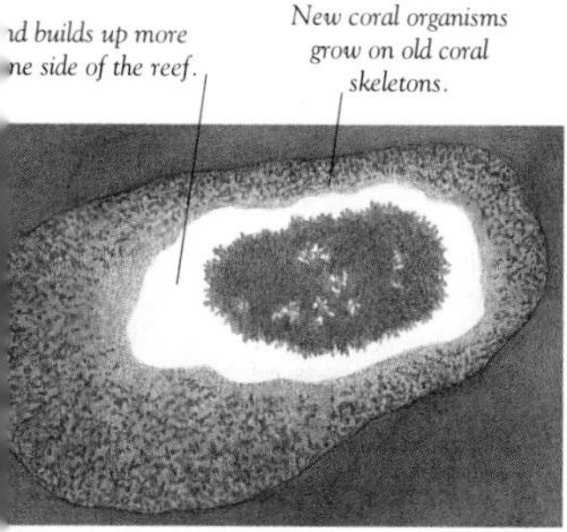
ıd builds up more
ne side of the reef.
New coral organisms
grow on old coral
skeletons.

ʀAL

nds such as the Maldives in the Indian
:an are known as coral. Tiny marine
ınisms called corals grow on submerged
k formations such as undersea
:anoes (seamounts) in warm, salty seas.
: coral grows slowly up to the ocean's
ace and, when the sea level drops,
ıtes a firm platform above sea level.

Formation of a Coral Atoll

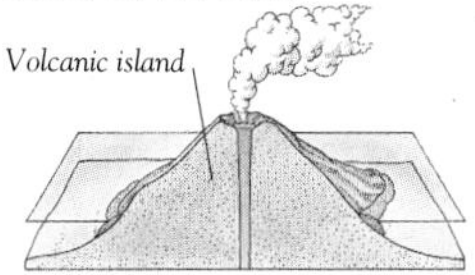
Volcanic island

1 A FRINGING REEF
Where a volcano has emerged from under the ocean, coral begins to grow on its fringes, around the base of the volcano.

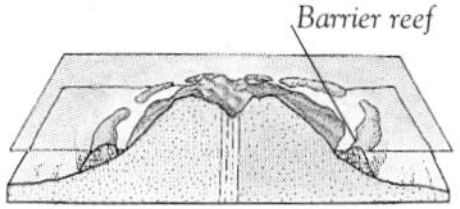
Barrier reef

2 A BARRIER REEF
When volcanic activity subsides, the peak erodes and sinks. The coral forms a reef around the edge of the volcano.

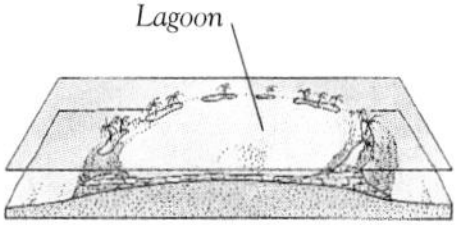
Lagoon

3 AN ATOLL FORMS
Eventually, the volcano sinks beneath the sea. A ring of coral known as an atoll remains on the surface.

COASTS

WHERE THE LAND meets the sea there is the coast. Coasts may be bordered by cliffs, dunes, or pebble beaches. There is a continual battle between sea and coast as rock is broken down by pounding waves, and sand carried about by wind. Some coasts retreat but new coast is always being create in other areas. Beaches alter their height and width with the seasons.

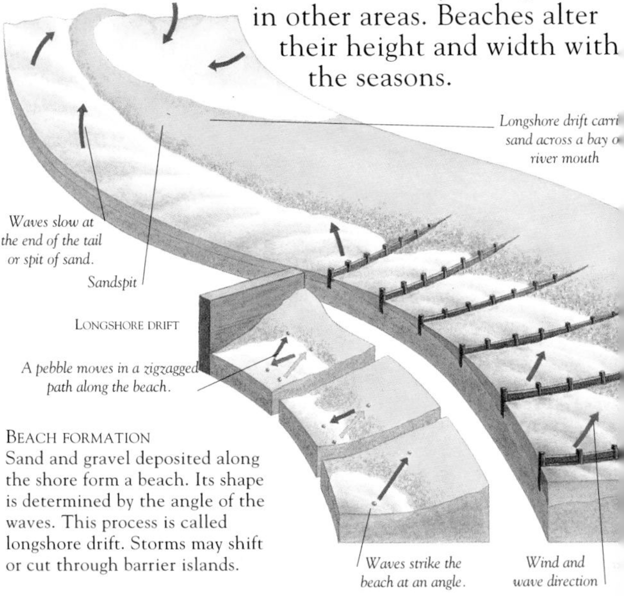
Longshore drift carri
sand across a bay o
river mouth
Waves slow at the end of the tail or spit of sand.
Sandspit
LONGSHORE DRIFT
A pebble moves in a zigzagged path along the beach.
Waves strike the beach at an angle.
Wind and wave direction

BEACH FORMATION
Sand and gravel deposited along the shore form a beach. Its shape is determined by the angle of the waves. This process is called longshore drift. Storms may shift or cut through barrier islands.

OAST FACTS

- The largest pleasure ·ach is Virginia Beach, ıth an area of 0 miles² (803 km²).
- The world has about 2,000 miles (504,000 n) of coastline.
- The highest sea cliffs e at Molokai, Hawaii. ney descend 3,300 ft ,010 m) to the sea.
- At Martha's Vineyard, ass., the cliffs retreat 5 ft (1.7 m) per year.

Groynes or fences built into the sea prevent longshore drift.
Sand builds up against the groin.
d and shingle

TYPES OF COAST

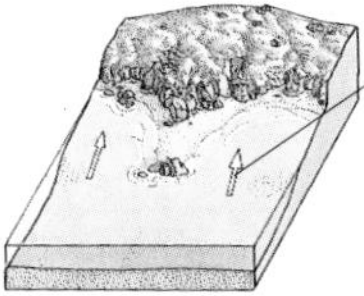

TOMBOLO
This type of coastline links an island to the shore by a strip of sand.

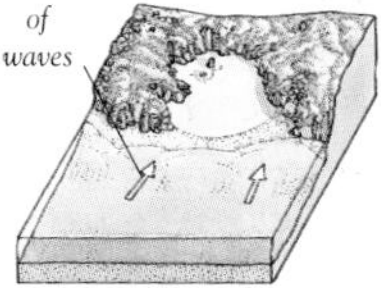

BARRIER BEACH
A lagoon forms behind a barrier of sand built by onshore waves.

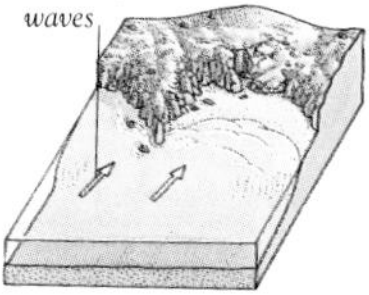

BAYHEAD BEACH
Waves striking a headland at an angle leave a protected arc of sand.

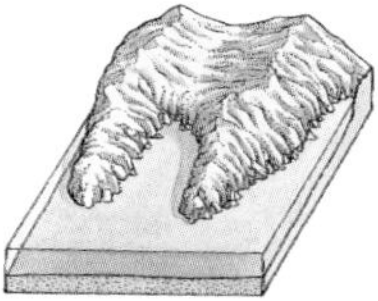

FJORD COASTLINE
A submerged glacial valley with steep sides forms a fjord coastline.

SEA STACKS
Waves, carrying sand and pebbles, gradually wear away a headland. First, a cave appears, which is enlarged to form an arch. Then, the arch falls, leaving an isolated stack.

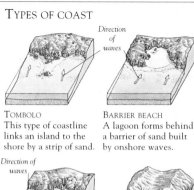

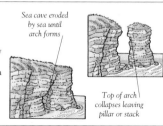

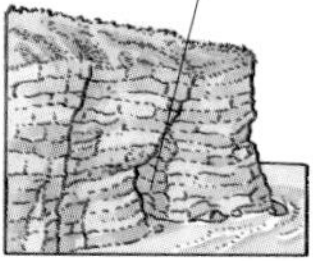

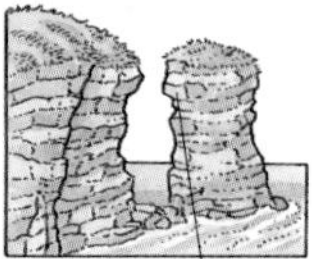

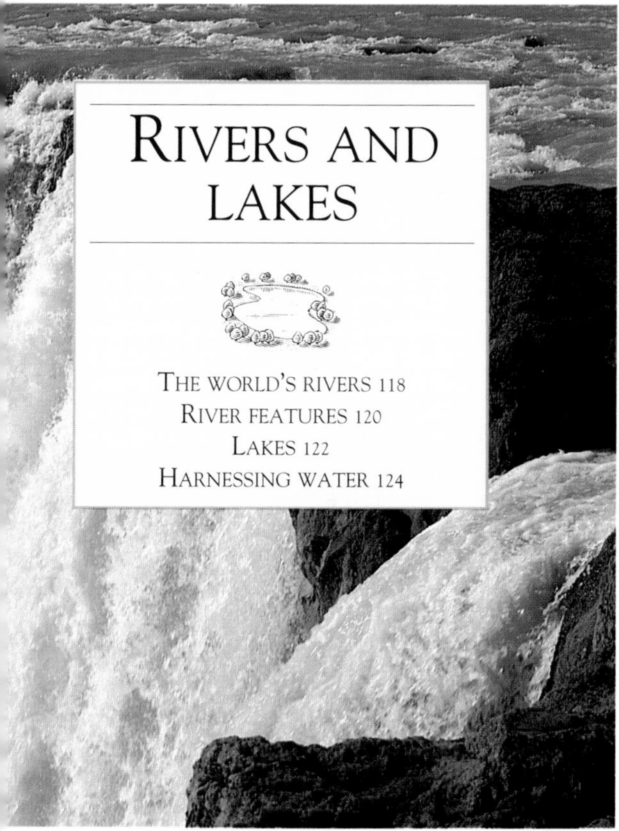

RIVERS AND LAKES

THE WORLD'S RIVERS

WHEREVER THEY occur, rivers are a key part of the Earth's water cycle. They carry snowmelt and rainwater from high areas down to the sea, filling up marshes and lakes where the land surface is uneven. In arid regions small rivers may be intermittent. They dry up and reappear after heavy rains or during an annual wet season.

PERENNIAL RIVER
In temperate and tropical areas a reliable supply of rainwater creat[e]s perennial rivers. Rivers such as t[he] Nile, in Africa, flow all year.

THE WATER CYCLE
Water is constantly cycling between land, sea, and air. The Sun's heat causes evaporation from seas, lakes, or rivers. Water vapor rises and cools to form tiny water droplets or ice particles seen as clouds. Precipitation falls on the oceans or is carried back by rivers or as groundwater.

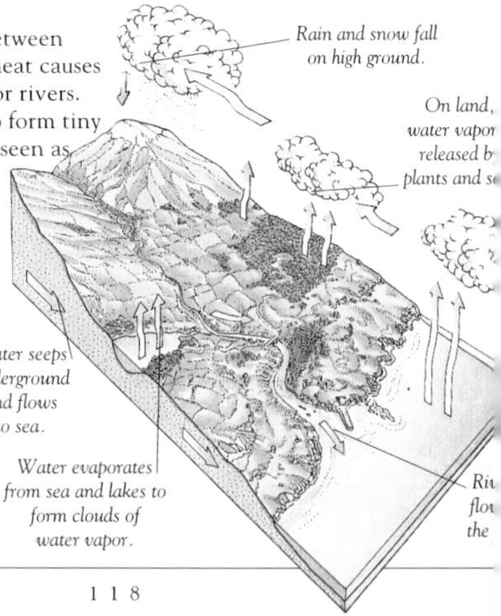

Rain and snow fall on high ground.
Water seeps underground and flows to sea.
Water evaporates from sea and lakes to form clouds of water vapor.

RIVER FACTS

- The Nile River is longer, but the Amazon carries more water.
- China's Yangtze carries 1,600,000 tons (tonnes) of silt a year.

ASONAL RIVER
is dry river bed belongs to a anish river. In the hot nmer many rivers dry up, but n will fill them up during the t winter season.

THE WORLD'S LONGEST RIVERS		
RIVER AND CONTINENT	LENGTH IN KM	LENGTH IN MILES
Nile, Africa	6,695	4,160
Amazon, S. America	6,437	4,000
Yangtze/Chang Jiang, Asia	6,379	3,964
Mississippi-Missouri, N. America	6,264	3,892
Ob-Irtysh, Asia	5,411	3,362
Yellow/Huang He, Asia	4,672	2,903
Congo/Zaire, Africa	4,662	2,897
Amur, Asia	4,416	2,744
Lena, Asia	4,400	2,734
Mackenzie-Peace, N. America	4,241	2,635

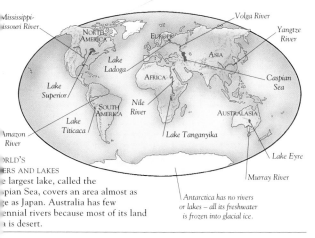

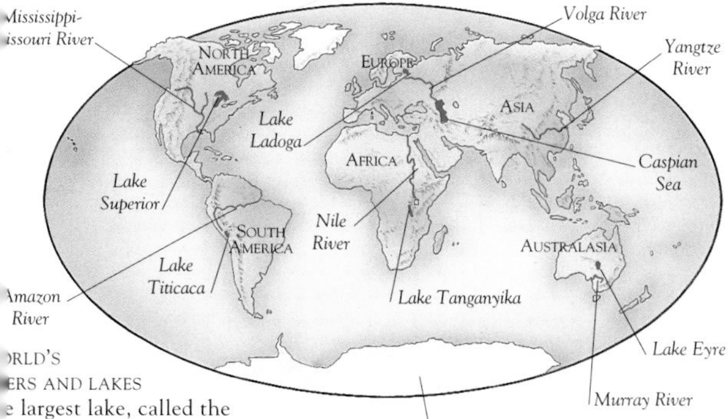

Antarctica has no rivers or lakes – all its freshwater is frozen into glacial ice.

RLD'S
ERS AND LAKES
e largest lake, called the pian Sea, covers an area almost as e as Japan. Australia has few ennial rivers because most of its land is desert.

RIVER FEATURES

FROM ITS SOURCE in the mountains, the snow or rainwater that fills a stream cuts a path through sediment and bedrock on its way to the sea. Streams join and form a river that flows more slowly, meandering across the land. A river may carry a large amount of sediment, which it deposits on its floodplain or in a delta.

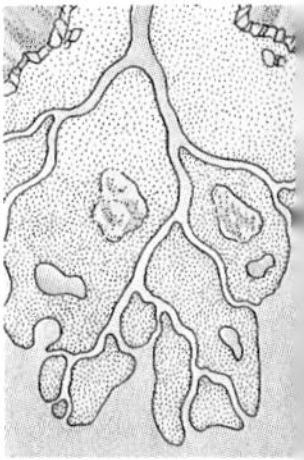

DELTA
At the river's mou
it sheds its sedimer
to form a broad fan
of swampy land.

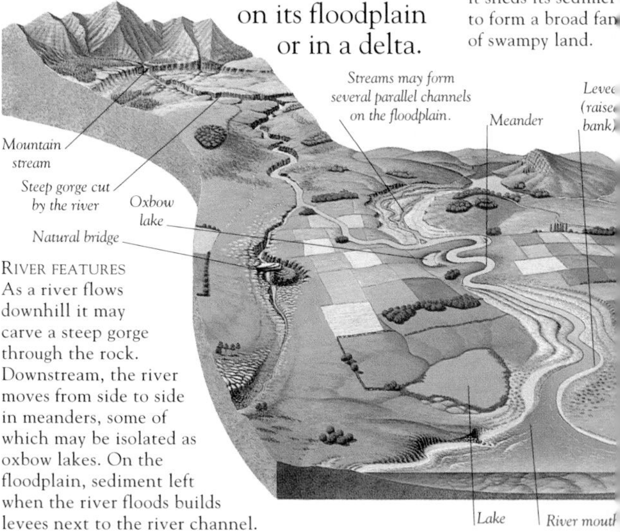

RIVER FEATURES
As a river flows downhill it may carve a steep gorge through the rock. Downstream, the river moves from side to side in meanders, some of which may be isolated as oxbow lakes. On the floodplain, sediment left when the river floods builds levees next to the river channel.

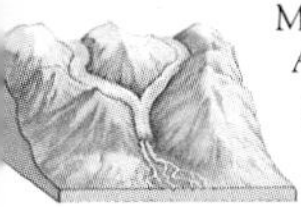

Meltwater

A river may begin its life in a glaciated part of the world. Melting ice and snow from a glacier feed mountain streams.

Overland flow

Rainwater running downhill gathers into small streams called tributaries, which join to form a river.

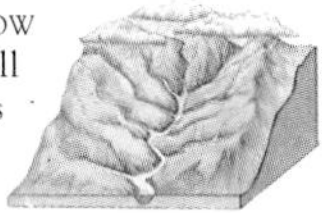

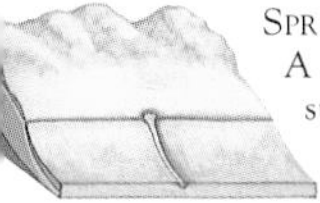

Spring

A rock layer called the aquifer stores rainwater. Springs form where streams cut into aquifer rock layers.

More river facts

- The Ganges and Brahmaputra delta, India, is the largest in the world. Its area is about 30,000 $miles^2$ (75,000 km^2).
- The widest waterfall is Khone Falls in Laos. They are 6.7 miles (10.8 km) wide.
- Each year rivers unload 20 billion tons (tonnes) of sediment into the sea.

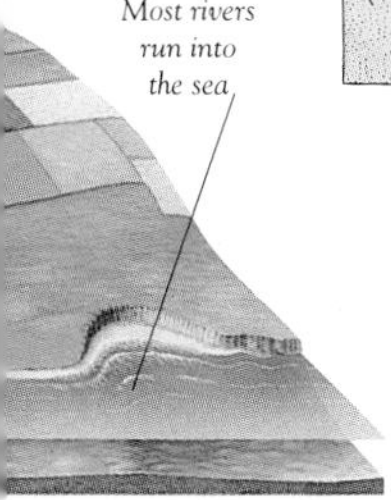

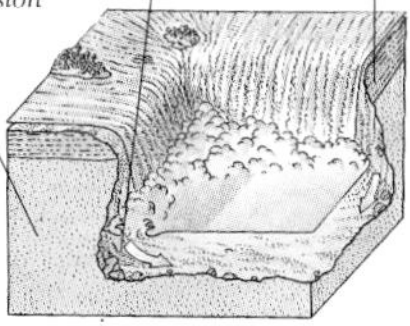

Waterfalls

A river flows swiftly near its source, cutting through soft rocks more easily than hard. A sheer face of hard rock is exposed where water plunges, undercutting the rock below.

The world's highest waterfalls

Waterfall and country	Height in meters	Height in feet
Angel Falls, Venezuela	979	3,212
Tugela Falls, S. Africa	853	2,799
Utgaard, Norway	800	2,625
Mongefossen, Norway	774	2,539
Yosemite Falls, US	739	2,425

LAKES

AN INLAND BODY of freshwater or brackish water, collected in a basin, is called a lake. In geological terms, lakes are short-lived; they can dry up or become clogged in a few thousand years. Lakes form when depressions resulting from lithosphere movement, erosion, or volcanic craters fill up with water. The Caspian Sea, the world's largest lake, and Lake Baikal, Siberia, the deepest lake, were both produced when shifts of the lithosphere cut off large arms of the sea.

SWAMP
The Everglades, Florida, are mangrove swamps. In warm climates, mangrove trees grow in the salty (brackish) water of mudd estuaries. The trees form islands in the mud.

TYPES OF LAKE

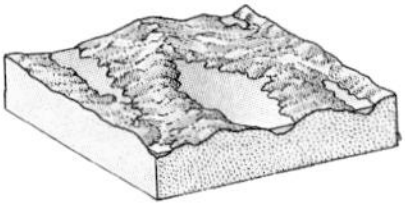

KETTLE LAKE
Ice left behind by a melting glacier may be surrounded by moraine. When the ice melts the depression forms a kettle lake.

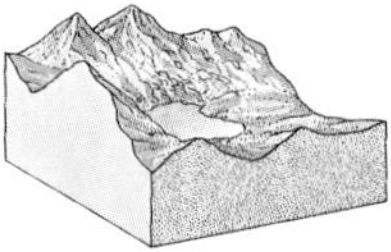

TARN
A circular mountain lake is known as a tarn. These lakes form in hollows worn by glacial erosion or blocked by ice debris.

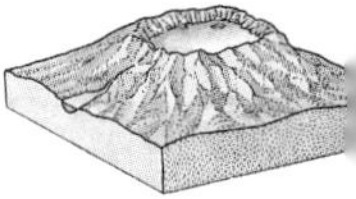

VOLCANIC LAKE
The craters of ancient volcanoes fill up with water and produce lake such as Crater Lake, Oregon.

VANISHING LAKES

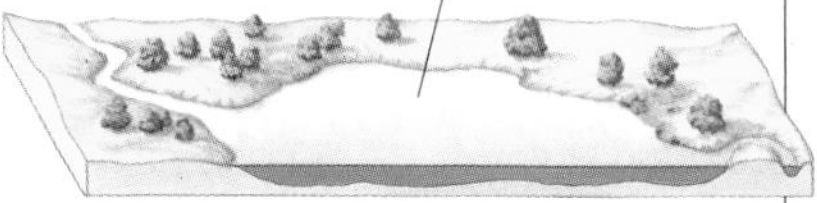
Silt and mud clog up lakes

SEDIMENT BUILDS
Lakes begin to fill up with sediment, washed into them by rivers. The mud and silt create a delta in the lake, which has areas of dry land.

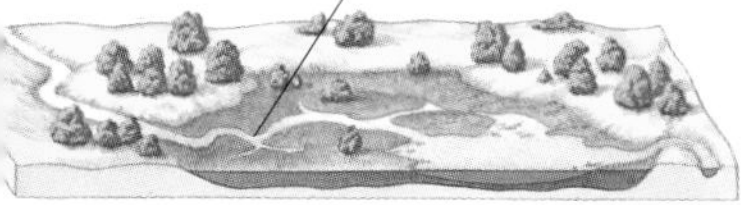
Channels become narrow

SWAMP FORMS
The lake area gets smaller and shallower. Islands of dry land fan out into the lake. Reeds grow, turning the lake into a swamp.

LAKE DISAPPEARS
Eventually, the lake area is colonized by plants, forming a wetland environment.

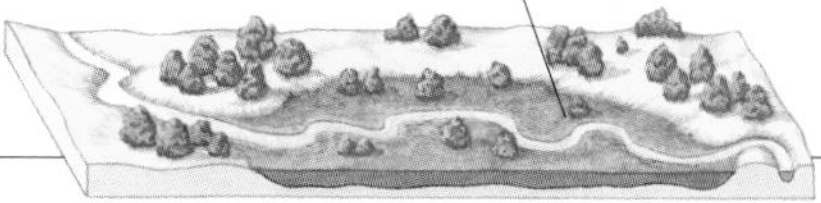
Plants grow in the sediment

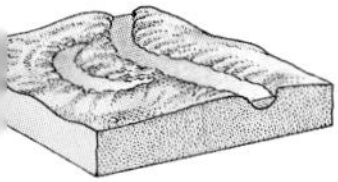

BOW LAKE
is curved lake appears en a river cuts off a ander loop. Eventually, isolated lake fills with ment and vegetation.

THE WORLD'S LARGEST LAKES AND INLAND SEAS

LAKE AND CONTINENT	AREA IN KM^2	AREA IN MILES^2
Caspian Sea, Asia/ Europe	370,980	143,236
Lake Superior, N. America	82,098	31,698
Lake Victoria, Africa	69,480	26,826
Lake Huron, N. America	59,566	22,999
Lake Michigan, N. America	57,754	22,299
Aral Sea, Asia	37,056	14,307
Lake Tanganyika, Africa	32,891	12,699
Lake Baikal, Asia	31,498	12,161
Great Bear Lake, N. America	31,197	12,045

HARNESSING WATER

HUMANS CANNOT survive without freshwater. It is necessary for human consumption, crops, and industry. Rainwater is stored in reservoirs and aquifers for later use. Riv flow can be channeled to crop irrigation networks and into canals that allo barge traffic. Water's moving energy can be harnessed in hydroelectri power stations to produc electricity. For domestic use water is cleansed, treated, and recycled.

IRRIGATION
Rice-terracing is a method of crop irrigation used in Indonesia. Growing rice requires a great deal of water. To make maximum use of rainfall, a system of channels carries water to the fields of rice. The fields are cut in terraces down the hillsides.

CANALS
During the Industrial Revolution in Britain in the 1800s, a network of waterways called canals was constructed. Goods could be transported by barges that were pulled along by horses. Aqueducts carry canals and their traffic over obstacles.

EANSING WATER

inwater in lakes or reservoirs must be ated with chemicals before it can be :d. In treatment plants, water passes ·ough filter beds, chlorine gas, and ıer chemicals to remove impurities. In pulated areas water goes through a nstant cleansing :le.

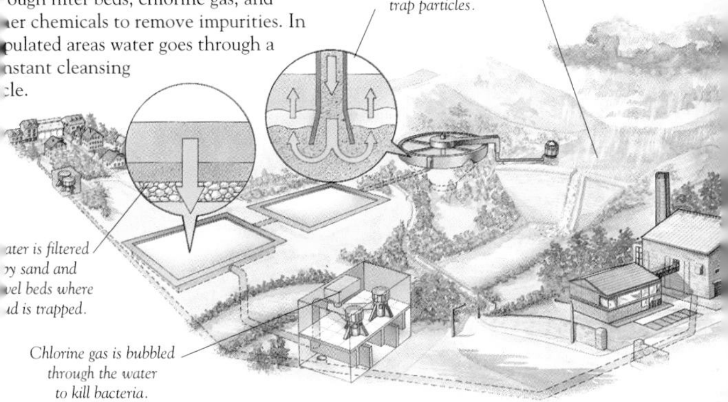

The chemicals aluminum sulfate and calcium hydroxide form a sticky substance to trap particles.

A dam holds back water in a reservoir.

ater is filtered y sand and vel beds where ıd is trapped.

Chlorine gas is bubbled through the water to kill bacteria.

ᴀMS

dammed river, like this e in California, can be :d to produce droelectric power. ater held in a ervoir by a dam ns a turbine that is ked to an electrical ıerator. In the same y, river water above ·aterfall can be ·erted to power bine generators.

DAM FACTS

- Water held by Volta Dam in Ghana could flood 3,282 miles^2 (8,500 km^2).
- A chain of dams being built along the Amazon could flood an area the size of England.
- There are more than 200 dams around the world that are over 492 ft (150 m) tall.

CLIMATE AND WEATHER

CLIMATE

TYPICAL LONG-TERM weather conditions for an area a known as its climate. The three broad climate zones are tropical, temperate, and polar. One factor that affects climate is distance from the equator (latitude). Different areas of the planet at the same latitude sha the same climate. The nearer the equator the warmer the climate; the nearer the poles the colder. Distance from the sea and altitude also affect climate.

TEMPERATE GRASSLAND
The temperate climates of North America and Northern Europe experience seasonal change but similar monthly rainfall all year.

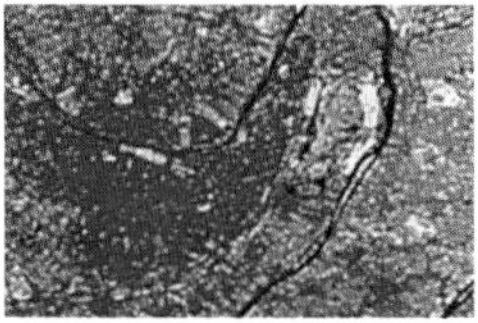

MICROCLIMATE
In a city, such as Paris, the weather may differ from that of outlying areas. Roads and buildings absorb heat to create a local or microclimate.

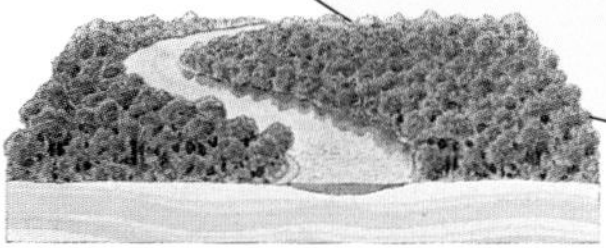

TROPICAL RAINFOREST
Regions of dense vegetation near the equator have a climate that is hot and wet all year round. The temperature stays constant at about 80–82°C (27–28°F).

POLAR REGIONS AND TUNDRA REGIONS

At the ice-covered poles, temperatures only rise above freezing for a few months of the year. The cold, dry tundra region surrounds the North Pole.

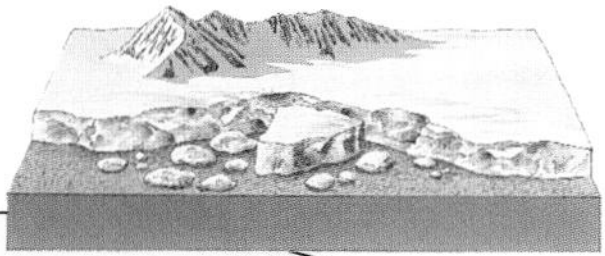

CLIMATE FACTS

- The temperature in the shade at al'Aziziyah, Libya, 136°F (58°C), is the highest recorded.
- Oymyakon in Siberia, the coldest inhabited place, can reach –90° F (–68°C).

MOUNTAIN REGIONS

The temperature falls the higher up a mountain you go. Trees and plants grow on the low slopes but little grows above the snowline.

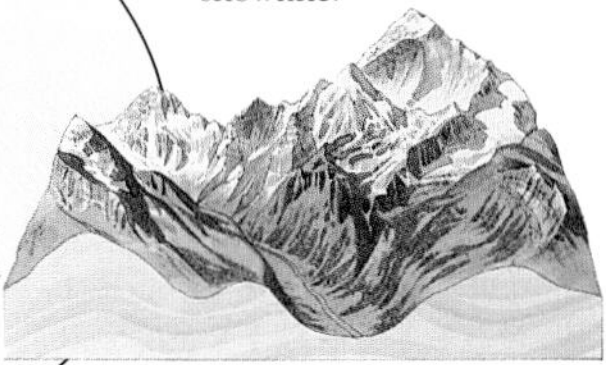

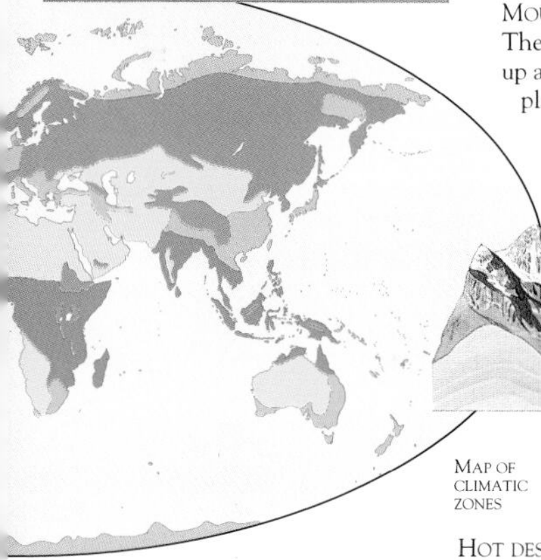

MAP OF CLIMATIC ZONES

KEY TO CLIMATIC ZONES

- Polar
- Tundra
- Mountain
- Temperate grassland
- Tropical rainforest
- Hot desert

HOT DESERT

Few animals and plants can live in the hot, dry conditions of the desert. The temperature can reach 100°F (38°C) and it may not rain for several years.

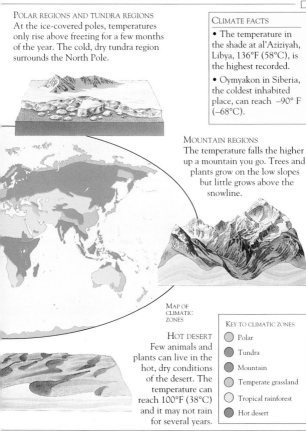

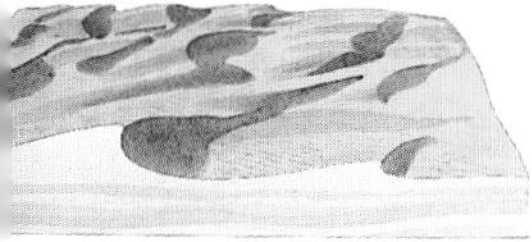

WIND AND WEATHER

WINDS CIRCULATE air around the planet. They carry warm air from the equator to the poles and cold air ir the opposite direction. This process balances the Earth's temperature. Some global winds (known as prevailing winds), such as polar easterlies and trade winds, are an important part of the world's weather systems.

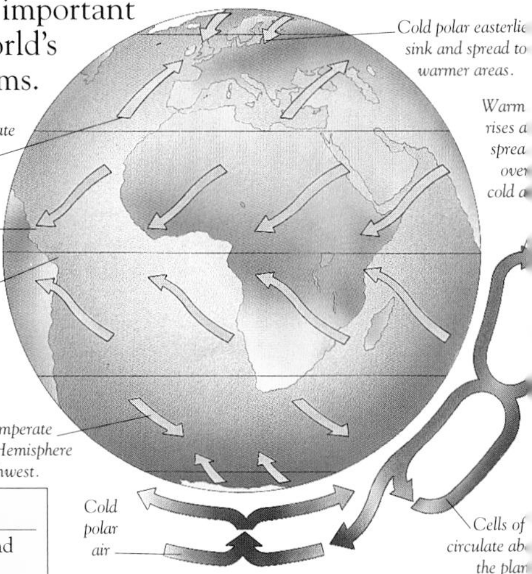

Cold polar easterlie
sink and spread to
warmer areas.
Warm
rises a
sprea
ove
cold a
Prevailing winds in temperate regions of the Northern Hemisphere blow from the southwest.
Trade winds
There is very little wind in the doldrums on the equator.
Hot air moves away from the equator toward the poles, where it cools.
Prevailing winds in temperate regions of the Southern Hemisphere blow from the northwest.
Cold polar air
Cells of
circulate ab
the plan

WIND FACTS

- In Antarctica wind can reach speeds of 200 mph (320 km/h).
- Highest wind speed recorded at ground level is 230 mph (371 km/h).

WINDS OF THE WORLD
Three prevailing winds blow around the planet at ground level, on either side of the equator. Trade win bring dry weather, westerly winds are damp and warm and polar easterlies carry dry, cold polar air.

'ORMATION OF A ROSSBY WAVE

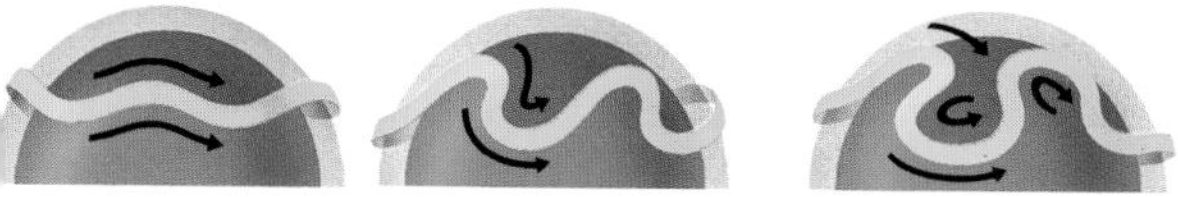

. SNAKING WIND
he Earth's rotation
auses curling, high
titude winds called
ossby waves.

DEEPENING WAVE
The wave deepens along the polar front. It forms a meander 1,250 miles (2,000 km) long.

DEVELOPED LOOPS
The curls become loops and the hot and cold air separate to produce swirling frontal storms.

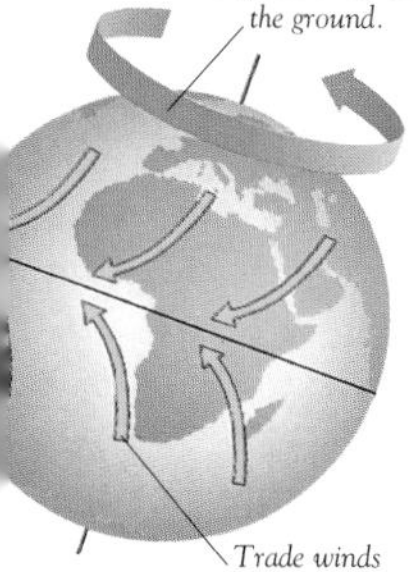
Earth's rotation deflects winds on the ground.
Trade winds near the equator

SEA BREEZES AND LAND BREEZES
On sunny days the land warms up during the day. Warm air rises from the land and cool air is drawn in from the sea. At night the land cools down quickly and cold air sinks out to sea.

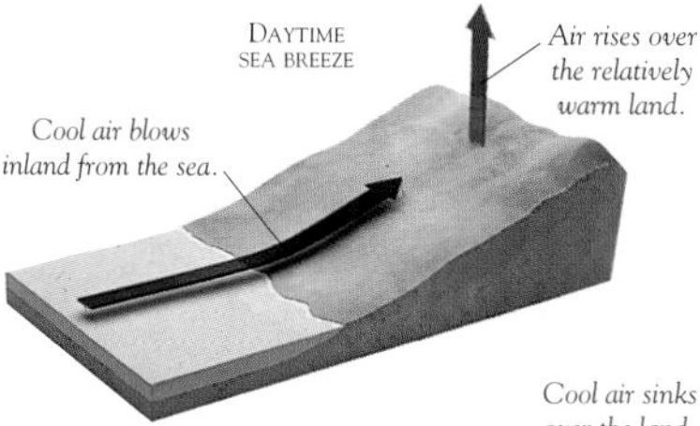
DAYTIME SEA BREEZE
Air rises over the relatively warm land.
Cool air blows inland from the sea.

Cool air sinks over the land.
NIGHTTIME LAND BREEZE
Air rises over the relatively warm sea.

DE WINDS
ne area either side of the
ator (the tropics) the
'ailing winds are called the
e winds. In the Northern
nisphere the winds blow
n the northeast, and in the
thern Hemisphere they
/ from the southeast.

WEATHER MAPPING

MILLIONS OF PEOPLE listen to the weather forecast each day. The forecast is compiled using data collected from all around the world and from weather satellites in space. Meteorologists study the movements of warm and cold air masses and the fronts where they meet. Using this information they plot weather charts and predict the coming weather.

SATELLITE IMAGES
From space, the Earth appears to be surrounded by swirling clouds. Afric desert has no cloud cove

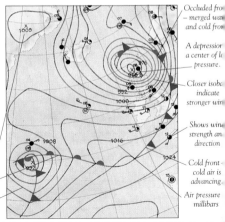

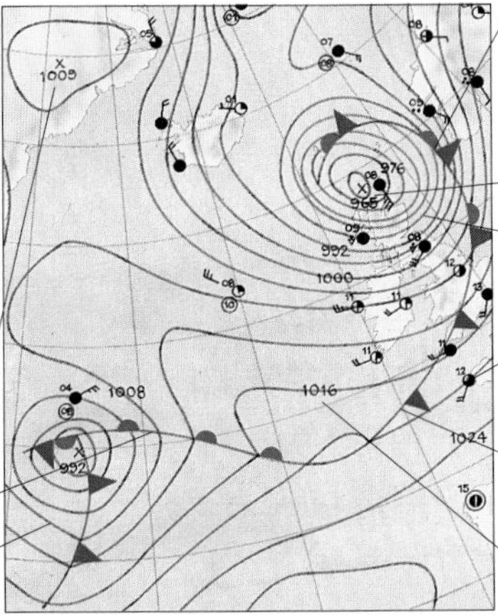

WEATHER MAP
A picture of the weather at a given time can be shown on a weather map, known as a synoptic chart. Standard symbols are used, such as lines to show fronts (where one body of air – an air mass – meets another).

ꟳORMATION OF A DEPRESSION

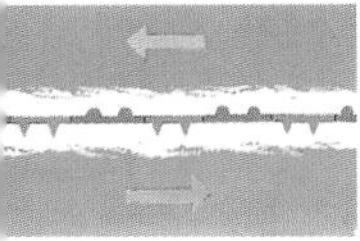

ꟷIR MASSES
ꟷir masses are a vast
ırea of wet or dry, warm
ır cold air. At the polar
ront, a warm air mass
ınd a cold one collide.

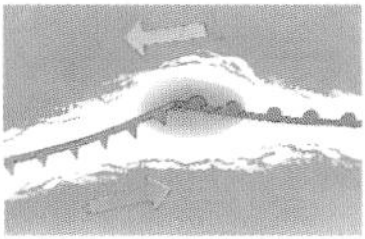

FORMING A BULGE
The warm tropical air mass pushes into the cold polar air along the polar front. The front begins to bulge.

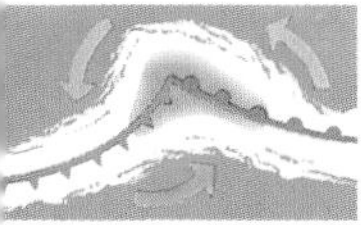

)IVIDING INTO TWO
'he Earth's rotation
pins the air masses.
Cold air pursues warm
ːr in a spiral formation.
'he polar front splits.

OCCLUDED FRONT
When the cold front catches up with the warm front, it pushes under the warm air. An occluded front results.

WEATHER FACTS

- A cold front can advance at up to 30 mph (50 km/h) and may overtake a warm front.
- The first television weather chart was broadcast in Britain on November 11, 1936.
- An air mass can cover an area as large as Brazil.

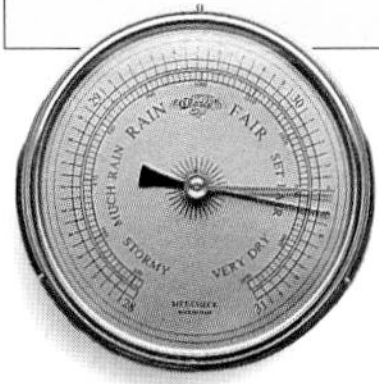

BAROMETER
The air around the Earth has mass and exerts pressure. A barometer measures air pressure in units called millibars.

ıPERATURE PEAKS
ughs in the temperature
ph show when ice sheets
anced to cover high
tude and land masses.
ese cold periods were
arated by interglacials
en the average
nperature rose and the
; sheets retreated.

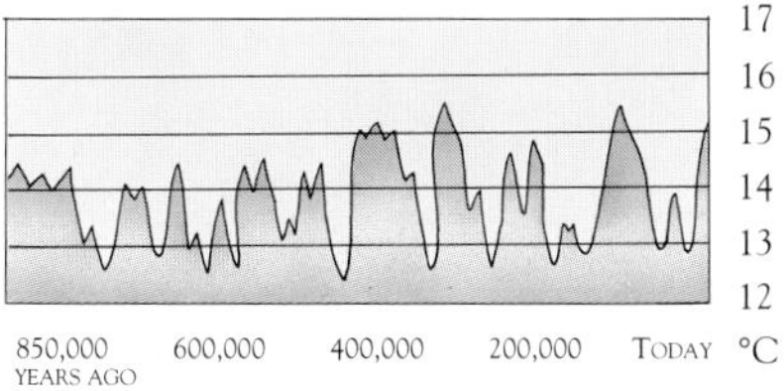

Clouds

AIR RISES as it warms, as it passes over mountains, or as it is pushed over air masses near the ground. Rising air cools, condenses, and forms clouds of water droplets. There are three cloud levels: cirrus form at the highest level, alto in the middle, and stratus at the lowest level.

FOGGY AIR
Clouds that form at ground level are known as fog. Fog, mixed with smok from burning fuels, produces smog. Earlier this century, London, Englan suffered from severe smog.

Cloud formation

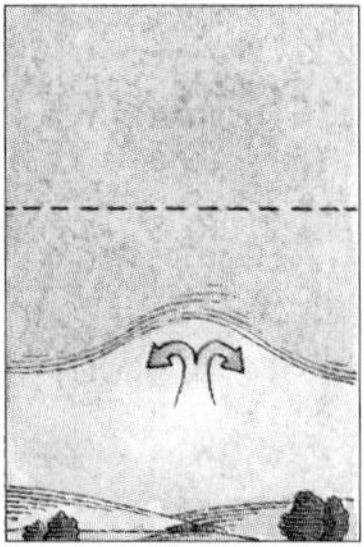

THE LAND WARMS
The Sun warms the land on a clear day. Air near the ground is warmed and rises.

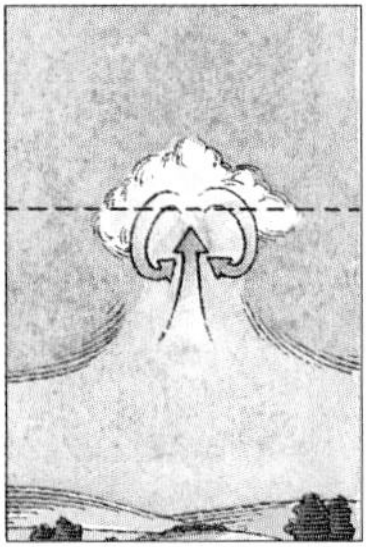

A CLOUD FORMS
As the warm air rises, it cools. The moisture it contains condenses and forms a cloud.

GROWING CLOUDS
Fleecy clouds appear in the sky. They get bigger and cool air circulates inside them.

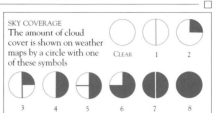

.OUD AND AIR FACTS

Cirrus are the highest ɔuds – they may reach ,370 ft (12,000 m).

Glider pilots and birds e pockets of rising air lled thermals to help em stay in the air.

SKY COVERAGE

The amount of cloud cover is shown on weather maps by a circle with one of these symbols

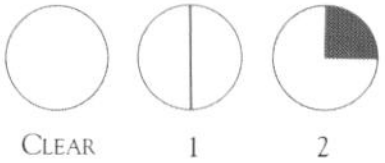

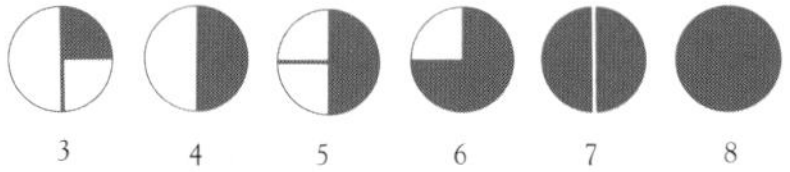

CLOUD TYPES

CIRRUS (A)
- wisps of cloud made of ice crystals
- about 39,372 ft (12, 000 m) high

CIRROCUMULUS (B)
- forms at about 29,529 ft (9,000 m)
- rippled ice crystal cloud

CUMULONIMBUS (C)
- dark, storm cloud with large vertical development

ALTOCUMULUS (D)
- layers or rolls of fluffy cloud

ALTOSTRATUS (E)
- gray or white sheet of cloud
- forms between 6,562 ft and 19,685 ft (2,000 m and 6,000 m)

STRATOCUMULUS (F)
- layer at the top of cumulus cloud

CUMULUS (G)
- large, heaped white fluffy cloud

NIMBOSTRATUS (H)
- low, rain cloud
- under 6,562 ft (2,000 m)

STRATUS (I)
- low-level, flat gray sheet of cloud

RAIN AND THUNDERSTORMS

EARTH'S WATER CYCLE relies on rain. Rain fills rivers and lakes and provides water for plants and animals. T water droplets in the air form rain when they gather i larger drops inside clouds. Raindrops can be moved around by air currents. L rain or snowfall may come from altostratu clouds, heavy rain fi stratus clouds, and torrential rains, or thunderstorms, fro cumulonimbus clou

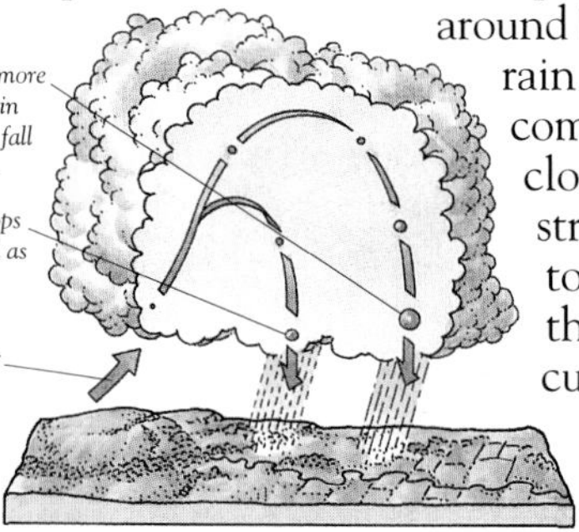

Droplets of more than 0.2 in (0.5 mm) fall as rain.
Smaller drops of water fall as drizzle.
Rising air

HOW RAIN FORMS
In warm regions, rising air currents agitate the water droplets in clouds until they join into raindrops. In temperate regions, ice crystals in the clouds above freezing level melt on their way down and form rain.

MONSOON
Seasonal winds called summer monsoons draw moist air inland, bringing rain to southern Asia. In winter, a cold, dry wind (winter monsoon) blows over land and out to sea.

RAIN FACTS

A record 73.62 in (1,870 mm) of rain fell in one month in 1861 in Cherrapunji, India.

Tutunendo, Colombia, the world's wettest place, has an annual rainfall of 463.4 in (11,770 mm).

LIGHTNING

Turbulence and collision of ice particles and water droplets inside a storm may cause electric charges. Positive charges gather at the top of the cloud and negative ones at the base. When the electricity is released it flashes between clouds or sparks to the ground and back.

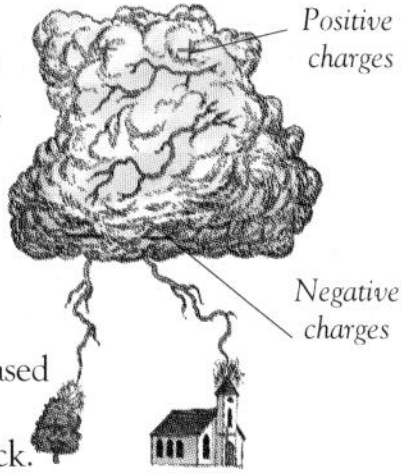

RAINBOWS

Sunlight striking raindrops is refracted, reflected by the backs of the droplets, and refracted again. This causes the white light to split into its seven constituent colors: red, orange, yellow, green, blue, indigo, and violet.

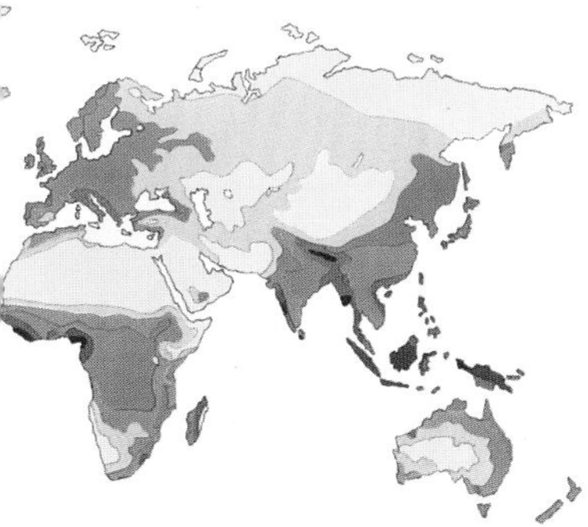

RAINFALL MAP

Around the world rainfall varies greatly. Warm seas in the tropics evaporate and bring lots of rain. Near the sea land is wetter but mountains may block rain.

KEY TO ANNUAL RAINFALL

- Less than 10 in (250 mm)
- 10–20 in (250–500 mm)
- 20–39 in (500–1,000 mm)
- 39–79 in (1,000–2,000 mm)
- 79–118 in (2,000–3,000 mm)
- More than 118 in (3,000 mm)

SNOW, HAIL, AND SLEET

WHEN THE WEATHER is very cold, snow, hail, or sleet leave a white coating on the landscape. Snow and hail result from water freezing in the clouds. Snow falls from the altostratus clouds, and hail from thunderstorm clouds. Sleet forms when partially frozen rainwater freezes completely as it touches any cold surface.

SNOWY WEATHER
In the European Alps the snow in the winter months does not melt because the ground temperature is low. Strong winds sometimes blow the snow into deep snowdrifts.

HOW SNOW FORMS
High up in the atmosphere, above the freezing level, water droplets in clouds form ice crystals, which collide and combine. As they fall, the ice crystals form snowflakes or hail.

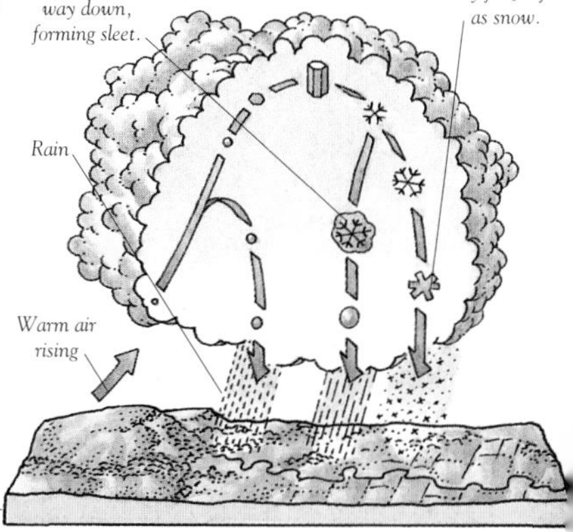
Some ice crystals melt on their way down, forming sleet.
Ice crystals whic stay frozen fall as snow.
Rain
Warm air rising

SNOW FACTS

- The largest recorded hailstone weighed 1.7 lb (765 g) and fell in Kansas in 1970.
- In 1921, 76 in (1,930 mm) of snow fell in Colorado in one day.

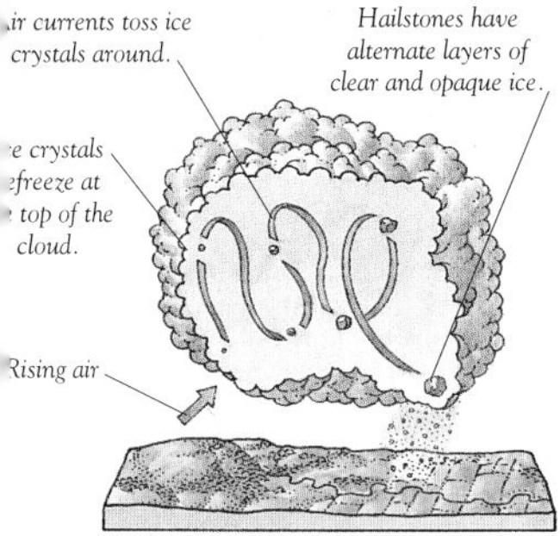
ir currents toss ice crystals around.
Hailstones have alternate layers of clear and opaque ice.
e crystals efreeze at top of the cloud.
Rising air

)W HAIL FORMS

cumulonimbus clouds above 6 miles) km) the temperature is freezing. Water)plets blown to the top of the cloud freeze. yers of ice build up around the hailstone it repeatedly melts and refreezes in its)tion through the cloud.

)ZEN RIVER

e Zanskar River in the Himalayas is frozen ing the winter months. The frozen river makes veling in the region easier, since local people 1 walk up and downstream on the ice. Under ice fish can survive in the unfrozen water. In nmer, the river is a fast-flowing torrent.

ICICLES

Spectacular ice shapes such as icicles form when water freezes in cold weather. Icicles grow as drips of melting snow or ice refreeze.

The outside of an icicle freezes before the inside.
Icicles often hang from leaking pipes.

HOAR FROST

Below freezing point, water vapor in the air freezes. It leaves spiky crystals of hoar frost on cold surfaces.

HURRICANES AND TORNADOES

FIERCE WINDS such as hurricanes and tornadoes occur when warm air masses encounter cold air masses. Winds reach high speeds and bring torrential rain and huge dark clouds. A tornado concentrates its havoc on a fairly narrow trail whereas a hurricane destroys a much larger area and can last for many days.

SATELLITE PHOTOGRAPH
Hurricanes can be tracked easily using hurricane hunter planes and weather satellites. The small centr eye develops as the hurricane reaches full intensity.

BEAUFORT SCALE	
NUMBER	DESCRIPTION
0	Calm, smoke rises straight up
1	Light air, smoke drifts gently
2	Light breeze, leaves rustle
3	Gentle breeze, flags flutter
4	Moderate wind, twigs move
5	Fresh wind, small trees sway
6	Strong wind, large branches move
7	Near gale, whole tree sways
8	Gale, difficult to walk in wind
9	Severe wind, slates and branches break
10	Storm, houses damaged, trees blown down
11	Severe storm, buildings seriously damaged
12	Hurricane, devastating damage

HURRICANE
Low-lying coasts are endangered by the high winds that generate large waves and also push water onshore as a destructive storm surge.

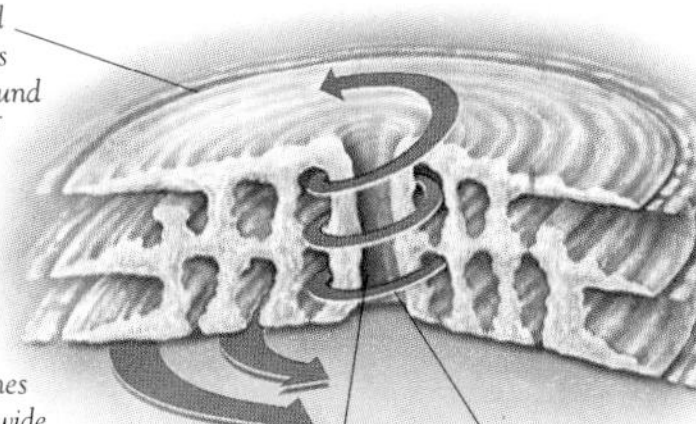

rus and
ostratus
uds around
edge of
storm.
Iurricanes
be as wide
480 miles
300 km).
Bands of wind
and rain spiral.
At the eye
the sky is
clear and
winds light.
Currents rise
and spread
outward.

HOW A HURRICANE FORMS
A cluster of tropical storms can become a hurricane. Bands of cumulonimbus and cumulus clouds spiral toward the center of the storm. Warm air rises and cools, building huge storm clouds that bring rain. At the center, or eye, of the hurricane, the pressure is low and the weather is calm.

ATERSPOUT
a tornado occurs over e ocean it is known as waterspout. Water is cked up in a column by nds reaching speeds of) mph (80 km/h), less an those of a tornado.

IURRICANE FACTS

- Winds up to 100 mph 160 km/h) have been ecorded in a hurricane.
- Hurricanes spin ounterclockwise north f the equator and clock-vise south of the quator.
- Waterspouts are isually between 64 and 330 ft 50–100 m) high.

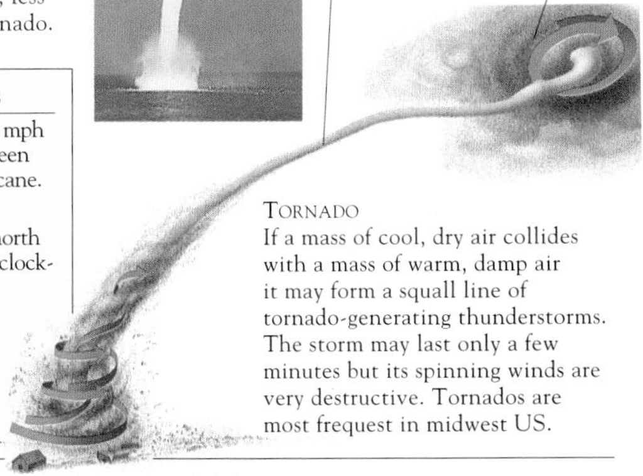

A violent updraft sucks up soil and debris.
Winds in the tornado may reach speeds of 200 mph (320 km/h).

TORNADO
If a mass of cool, dry air collides with a mass of warm, damp air it may form a squall line of tornado-generating thunderstorms. The storm may last only a few minutes but its spinning winds are very destructive. Tornados are most frequest in midwest US.

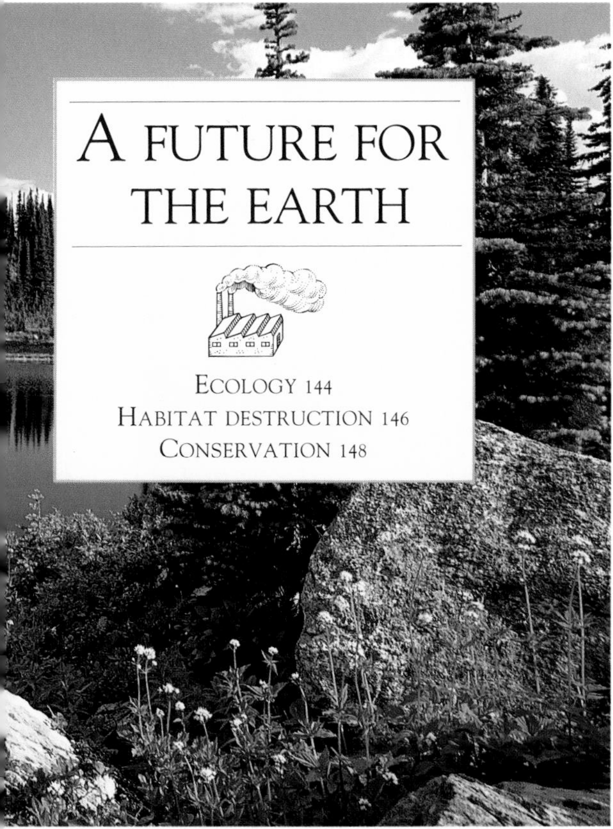

A FUTURE FOR THE EARTH

ECOLOGY

THE STUDY OF the relationships between animals and plants, and between them and their environment, is called ecology. Ecology explains how individual species fit into the natural world. Ecologists study how organisms obtain food and materials to survive and the effect this has on the environment and on other organisms. Ernst Haekel, a German biologist, first used the term "ecology" in 1866.

ECOSYSTEMS
Several communities of living things their physical surroundings, and their climate make up an ecosystem. Beach forest, and ocean communities are examples of ecosystems.

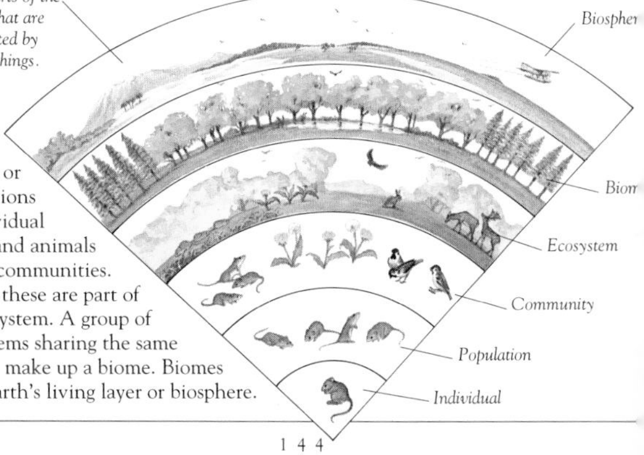

LIFE LAYERS
Groups or populations of individual plants and animals live in communities. In turn these are part of an ecosystem. A group of ecosystems sharing the same climate make up a biome. Biomes form Earth's living layer or biosphere.

·OD WEB

eding relationships
tween organisms in
. ecosystem can be
mplex. Energy, as
ɔd, is transferred
ɔng a chain from
ınts to herbivores
carnivores. Chains
terconnect to build a
ɔd web.

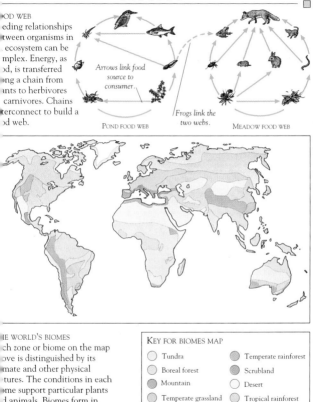

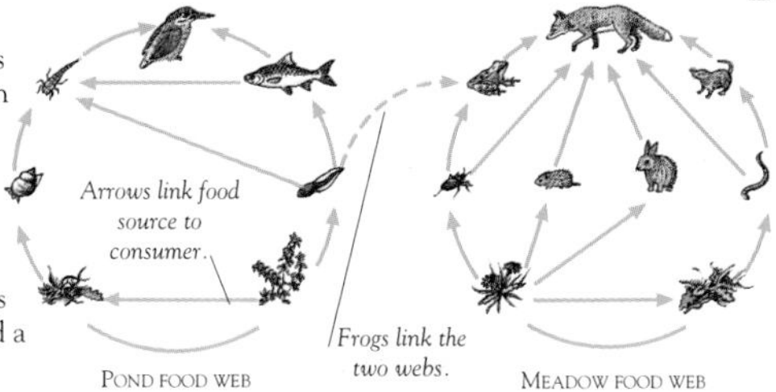
Arrows link food source to consumer.
Frogs link the two webs.
POND FOOD WEB
MEADOW FOOD WEB

IE WORLD'S BIOMES

ch zone or biome on the map
ɔve is distinguished by its
mate and other physical
tures. The conditions in each
ɔme support particular plants
d animals. Biomes form in
nds roughly parallel to the
es of latitude.

KEY FOR BIOMES MAP

- Tundra
- Boreal forest
- Mountain
- Temperate grassland
- Temperate forest
- Temperate rainforest
- Scrubland
- Desert
- Tropical rainforest
- Savanna

HABITAT DESTRUCTION

AS HUMANS SEEK a more comfortable life the Earth's resources are being used up or destroyed. Fossil fuels are burned to provide energy and waste is dumped which pollutes land, sea, and air. Toxins that do not break down are left to poison the planet. Cultivating more and more land has led to the loss of habitats like the rainforest, endangering its many unique species. We need to protect the Earth – our own habitat – for our own survival.

DEFORESTATION
Forests are destroyed as people clear land for their animals, to grow crops, or to sell the wood. Deforestation leads to habitat and wildlife loss.

Oil spilled at sea is washed up on the beach.

WATER POLLUTION
Once in the water cycle, pollutants such as chemical waste, gasoline, or oil can contaminate surface and ground water. Pollution of oceans and beaches kills animals and plants and poisons their habitats.

DESERTIFICATION
Deserts are growing because of climate change and overgrazing, which cause soil erosion.

CID RAIN
ees in Europe and
orth America suffer
mage by acid rain.
lfur dioxide is
eased into the air
ien fossil fuels are
rned. This mixes
h rainwater to
oduce a weak
furic acid, which
ls areas of forest.

POLLUTION FACTS

- The temperature on Earth could rise by 7°F (4°C) by the year 2050.
- A 3.3 ft (1 m) rise in sea level could flood 310,694 miles (500,000 km) of coastline.
- Each CFC molecule can destroy 100,000 molecules of ozone.
- An area of rainforest the size of a soccer field is destroyed every second.

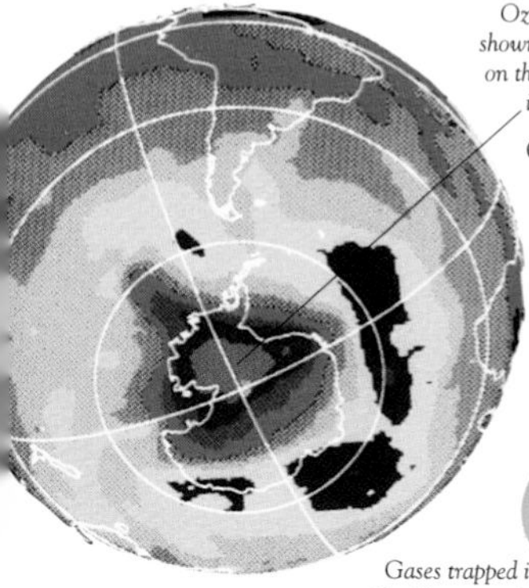

Ozone hole shown in purple on this satellite image

OZONE HOLE

The ozone layer blocks out most of the Sun's harmful ultraviolet radiation. Leakage of CFCs (chlorofluorocarbons), used in refrigerators, packaging, and aerosols, may be causing a hole in the ozone layer over Antarctica.

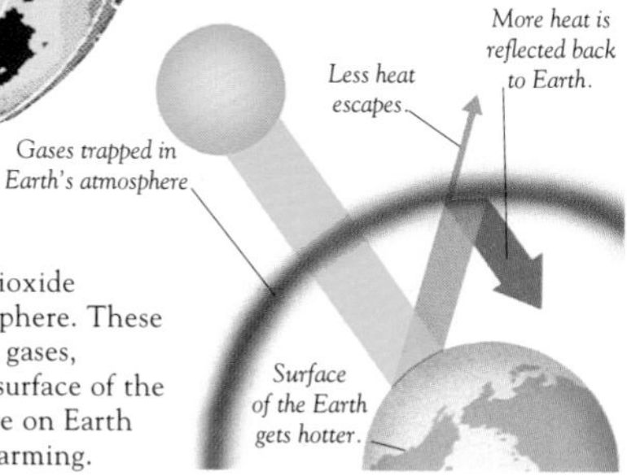
More heat is reflected back to Earth.
Less heat escapes.
Gases trapped in Earth's atmosphere
Surface of the Earth gets hotter.

OBAL WARMING
luting gases such as carbon dioxide
umulate in the Earth's atmosphere. These
es, also known as greenhouse gases,
vent heat escaping from the surface of the
net and cause the temperature on Earth
ise, a process called global warming.

CONSERVATION

ENDANGERED wildlife can benefit from our attempts to protect the environment. Animals and plants need their habitats preserved. Some endangered animal and plant species need the protection provided only by zoos and botanical gardens. Less waste and less pollution will also protect our unique planet.

CAPTIVE BREEDING
Endangered species such as pandas are encouraged to breed in captivity. This practice helps maintain and increase animal numbers.

Foxes are a familiar sight in gardens.

ADAPTING TO CHANGE
Animals such as foxes and raccoons have adapted to increased urbanization. Many now live in parks and gardens and scavenge for food in trash cans.

WILDLIFE RESERVES
One species becomes extinct every day through hunting or habitat destruction. However, in some parts of the world animals such as rhinos, zebras, and elephants are protected in wildlife reserves where they can live and breed in a protected and natural environment.

RECYCLING TRASH

An average family throws away two tons (tonnes) of trash every year. A large proportion of household waste such as organic matter, glass, paper, metal, and some plastics, all have good recycling potential. The remaining trash is still deposited in a diminishing number of land fills.

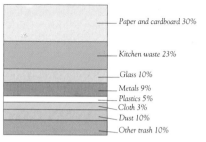

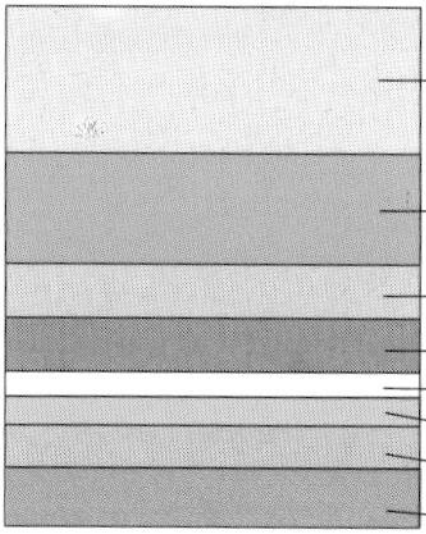

LAST WILDERNESS

Frozen Antarctica is a nearly deserted land, but the sea around it is teeming with life. Toxic dumping and mining are banned by an international convention in order to preserve Antarctica as Earth's last wilderness.

CONSERVATION FACTS

- Ingredients in some medicines, chocolate, and chewing gum were originally discovered in the rainforest.
- The bald eagle is no longer endangered because of breeding programs and protection laws.
- California plans to generate 10 percent of its electricity using wind turbines by the year 2000.

TURE ENERGY

ssil fuels that are burned to provide electricity will entually run out. The Sun is a possible alternative ergy source. Electric cars with solar panels are not practical but solar-powered telephones are eady in use in some sunnier ts of the rld.

Glossary

Abyssal plain
A sediment-covered plain on the ocean floor.
Acid rain
Rain that is acidic due to pollution by sulfur dioxide and other gases.
Anticyclone
Area of high atmospheric pressure.
Asthenosphere
Semimolten mantle layer that lies beneath the lithosphere.
Atmosphere
Envelope of air that surrounds the Earth. It can be divided into four layers: troposphere, stratosphere, mesophere, and thermosphere.
Atoll
A circular line of coral islands around a lagoon.
Avalanche
A rapid movement of unstable snow, ice, or rock down a mountain slope.
Barometer
An instrument that measures atmospheric pressure.
Basalt
The most common extrusive igneous rock. It is fine-grained and varies from dark gray to black in color.
Beach
Strip of loose sediment that lies between the cliffs or dunes and the lowest tides.
Beaufort scale
A twelve-step scale using observations of ocean waves, flags, and smoke to gauge wind velocity.
Biosphere
Layer where life is found on Earth.
Black smoker
Chimneylike vent on a rift valley floor from which black, mineral-rich hot water gushes.
Breccia
Sedimentary rock composed of coarse, angular rock fragments held together by a mineral cement.
Canyon
A deep valley with almost vertical sides, eroded by a river.
CFC
Chlorofluorocarbon containing carbon, chlorine, and fluorine. It is used in aerosol sprays, packaging, and refrigerators.
Cirrus cloud
An ice-crystal cloud that forms in the upper troposphere.
Cleavage
A well-defined plane along which a mineral tends to break, related to weakness in the atomic structure of the mineral.
Climate
The average weather conditions for a region over a long period of time.
Contact metamorphism
Baking and mineral change in rock near a magma chamber or in contact with dikes, sills, or lava flows.
Continental drift
Theory suggesting that the Earth's continents have moved (drifted) relative to one another.

CORE
The Earth's center sphere of nickel-iron, with a molten outer shell and solid inner part.

CRUST
The upper rock zone of the lithosphere.

CUMULONIMBUS CLOUD
Tall convective cloud that often brings rain.

CUMULUS CLOUD
Fluffy cloud usually accompanying fine weather.

CYCLONE
An area of low atmospheric pressure. Also a hurricane.

DEFORESTATION
Cutting down trees and clearing forests for crops, timber, and grazing land.

DESERTIFICATION
Creation of deserts by soil erosion through over-grazing, over-cultivation, or over-population.

ECOLOGY
Study of interactions between living organisms and their environment.

ECOSYSTEM
Communities of animals and plants dependent on each other and their environment for survival.

ENVIRONMENT
The life forms and air, water, and soil conditions of an area.

EPICENTER
Point on the Earth's surface directly above the focus of an earthquake.

EROSION
The removal of soil and weathered rock by wind, flowing water, or glaciers.

ESTUARY
The drowned mouth of a river where freshwater mixes with ocean water.

EXTRUSIVE IGNEOUS ROCK
Rock formed from lava erupted from volcanoes.

FIRN
Old, dense, granular compacted snow.

FISSURE ERUPTION
Volcanic eruption from a crack or linear vent in the ground.

FJORD
A former glacial valley with steep sides and a U-shaped profile, that is now occupied by sea water.

FOCUS
Underground location where an earthquake originates.

FOSSIL
Remains or imprints of animals and plants preserved in rock.

FRACTURE
A break in a mineral or rock that is not related to its atomic structure.

FRONT
Area where two different air masses collide.

GEYSER
A fountain of hot water or steam heated by volcanic activity.

GLACIER
Mass of ice on land that flows downhill under its own weight.

GLOBAL WARMING (GREENHOUSE EFFECT)
An increase in the global temperature as a result of heat being trapped in the atmosphere by gases such as carbon dioxide.

GRANITE
A coarse-grained intrusive igneous rock.

GUYOT
Flat-topped submarine (undersea) mountain.

HABITAT
The environment in which an animal or

plant lives.

HIGH
An area of high atmospheric pressure

HOAR FROST
Water vapor from fog that crystallizes as ice on rough, cold surfaces.

HURRICANE
Violent tropical storm, with high winds and torrential rain.

IGNEOUS ROCK
Rock formed when magma or lava solidifies.

INTRUSIVE IGNEOUS ROCK
Rocks resulting from injection of lava into existing rocks.

ISOBAR
Line on a weather chart joining points with equal air pressure.

LAVA
Magma erupted from fissures and volcanoes.

LIGNITE
Soft, woody coal formed by the burial of peat.

LIMESTONE
Sedimentary rock, mostly calcium carbonate.

LITHOSPHERE
The outer layer of the Earth which contains the crust and upper part of the mantle.

LONGSHORE DRIFT
The movement of sand and sediment along a beach by waves moving obliquely to the shore.

LOW
An area of low atmospheric pressure.

MAGMA
Molten rock material under the Earth's surface.

MANTLE
The thick layer between the core and the crust of the Earth.

MARBLE
A metamorphic rock, formed when limestone or another carbonate rock is changed by heat and/or pressure.

MERCALLI SCALE
A 12-step scale that rates an earthquake by its destructiveness.

METAMORPHIC ROCK
Rock formed by the effect of pressure and heat on existing rock.

METEOROLOGY
The scientific study of weather.

MIDOCEANIC RIDGE
The huge mountain range that runs through all ocean basins.

MINERAL
A naturally occurring substance with a constant chemical composition

ORBIT
The path taken by a planet as it travels around the Sun, or by a moon or spacecraft which travels around a planet.

ORE MINERAL
Mineral that contains enough metal or nonmetal to make its removal profitable.

OZONE LAYER
Layer in the upper atmosphere containing ozone, a gas that absorbs the Sun's ultraviolet rays

PALEONTOLOGY
Scientific study of fossils.

PEAT
Dark soil formed by the partial decomposition of vegetation in wet areas of marsh or swamp.

PERMEABLE ROCK
Rock that allows water and other liquids, such as oil, to pass through it.

PLATE TECTONICS
A theory suggesting the lithosphere is made of rigid plates that move relative to each other.

POLLUTION
Gases, liquids, or solids, largely released ir deposited by humans, that contaminate the environment.
PRECIPITATION
Rain, sleet, hail, and snow that fall to the ground from clouds.
PREVAILING WIND
The usual or common wind direction for an area.
RAINBOW
Colored arc seen in the sky, formed when sunlight splits into the colors of the spectrum.
REGIONAL METAMORPHISM
Large-scale change of rock that results from plate collision and mountain building.
RICHTER SCALE
A measure of earthquake energy based on the amplitude of surface waves recorded on seismographs.
RIFT (GRABEN)
Sinking of a strip of the lithosphere between two faults.
ROCK
Solid mass made of one or more minerals.
SAVANNA
Grassland area at the edge of the tropics which has seasonal rain.
SCREE
Mass of boulders and smaller fragments that accumulate at the bottom of cliffs and mountain slopes.
SEISMIC WAVE
A shock wave from an earthquake, measured using a seismometer.
SEISMOLOGY
The study of earthquakes and earth structure.
SMOG
A mixture of smoke and fog.
SOLAR SYSTEM
The Sun and the planets, meteors, comets, moons, and asteroids orbiting the Sun.
SPREADING RIDGE
Submarine mountains where two plates are moving apart and new crust is being created.
STALACTITE
Conical mineral deposit hanging from a cave roof.
STALAGMITE
Conical mineral deposit that builds up from the cave floor.
STRATUS CLOUD
Low-lying, layered cloud.
STREAK
The color of a mineral in its powdered form.
SUBDUCTION ZONE
Plane along which oceanic lithosphere sinks beneath an opposing plate.
TORNADO
Violent thunderstorm producing a destructive funnel cloud underneath.
TRADE WINDS
Winds that blow from high pressure regions of subtropical belts toward areas of low pressure at the equator.
TSUNAMI
Destructive sea waves caused by earthquakes under the ocean.
VOLCANO
A vent in the lithosphere through which magma erupts as lava.
WEATHER
Atmospheric conditions at a particular time and place.
WEATHERING
Physical and chemical processes that break down rocks on the Earth's surface.

Index

Acknowledgments

Dorling Kindersley would like to thank:
DK Cartography for the maps, Hilary Bird for the index, and Robert Graham and Connie Mersel for editorial assistance.

Photographs by:
J. Stevenson, C. Keates, A. von Einsiedel, H. Taylor, A. Crawford, G. Kevin, D. King, S. Shott, K. Shone.

Illustrations by:
J. Temperton, J. Woodcock, N. Hall, R. Ward, E. Fleury, B. Donohoe, D. Wright, C. Salmon, B. Delf, P. Williams, S. Quigley, R. Shackell, R. Lindsay, P. Bull, P. Visscher, R. Blakeley, R. Lewis, L. Corbella, D. Woodward, C. Rose, N. Loates, G. Tomlin.

Picture credits: t=top b=bottom c=center l=left r=right
AKG London: Erich Lessing / Galleria dell'Accademia 68bl. Bridgeman Art Library: British Museum 77bl; Christie's, London 48-49t. British Coal: 79tl. British Crown: 72tr. Bruce Coleman: G. Cubitt 77br; A. Davies 55bl, br; Dr. M. P. Kahl 149cl; H. Lange 74-75; W Lawler 144t; Dr. John MacKinnon 148tr; M. Timothy O'Keefe 83b; Fritz Prenzel 57tl; Andy Purcell 139tr; Gunter Ziesler 148br. Ecoscene: A. Brown 52tl, 139br, 142-143; R. Glover 136bc; Pat Groves 119t; Sally Morgan 52bl, 62tr; Tweedie 68tl; P. Ward 146br. Mary Evans Picture Library: 32bl, 44c. Frank Lane Picture Agency: 36bl, H. Hoslinger 141c; S. Jonasson 41cl; S. McCutcheon 42-43, 49bl; National Park Service 30-31. GeoScience Features Picture Library: 36tl, 39tr. Robert Harding Picture Library: 50-51, 63tl, 93tl; D. Hughes 90tl; Krafft 38br; R. Rainford 86tl. Hulton Deuts Collection: 77tl, 99bl, 124b, 134tr. Image Bank: 86bl, 88br; L. Brown 126-127; G. Champlong 122t; T. Madison 61tl; B. Rous 137c. Mountain Camera: J. Cleare 32tl. NASA: 10-11, 140tr. National History Museum: 65bc, 66cl, 67bl. Oxford Scientifi Films: Ben Osborne 146bl. Planet Earth: J. Downer 124t; J. Fawcett 40tl; R. Hessler 41t C. Huxley 2tr, 102-103; J. Lithgoe 98tr; B. Merdsoy 94-95; W. M. Smithey 93bl; N. Ta 90bl. Rex Features: A. Fernandez 37t; SIPA Press 46tl. Science Museum, London: 149b. Science Photo Library: D. Allan 92t; Tony Craddock 84-85; Dan Farber 140br; S. Frase 139bl; J. Heseltine 54tl; D. Pellegrini 100tr; Dr. Morley Read 146tl; NASA 112; F. Sauze 106tr; US Geological Survey 22-23. Solarfilma: 38tr. Frank Spooner Pictures: Barr/Liaison 36br; Garties/LN 39tl; Vitti/Gamma Liaison. Tony Stone Images: 60t, 62br, 80tr, 82tr; T. Braise 125b. Tony Waltham: 32cl, 34tl, 116-117. L. White, 118t. Zefa: 48br, 88cl, 138t, 147tc.

Every effort has been made to trace the copyrigh holders and we apologize in advance for any unintentional omissions. We would be pleased t insert the appropriate acknowledgments in any subsequent edition of this publication.